Bryan Sykes, Professor of Human Genetics at the University of Oxford, has had a remarkable scientific career. After undertaking medical research into the causes of inherited bone disease, he set out to discover if DNA, the genetic material, could possibly survive in ancient bones. It did, and his was the first report on the recovery of ancient DNA from archaeological bone, published in the journal *Nature* in 1989. Since then Professor Sykes has been called in as a leading international authority to examine several high-profile cases, such as the Ice Man, Cheddar Man and the many individuals claiming to be surviving members of the Russian Royal Family.

Alongside this, he and his research team have over the last ten years compiled by far the most complete DNA family tree of our species yet seen.

As Professor Sykes has always emphasized the importance of the individual in shaping our genetic world, he founded his own company that offers services to allow people from all over the world to explore their genetic roots. Oxford Ancestors Ltd (www.oxfordancesters.com) gives men and women the opportunity to find their place in the family tree of the entire human race through DNA based analysis and a chance to find their ancient ancestors.

Acclaim for
THE SEVEN DAUGHTERS OF EVE

'This is a wonderful tale of archaeology and genetics that should be read by anyone concerned with what we are . . . This is a terrific book, written with humour and a humanity that shames the racist strife lurking in modern Europe'
Sunday Times

'An engrossing, bubbly read, a boy's own adventure in scientific story-telling that fairly bounces along . . . there is no doubt he has produced some breathtaking work . . . a thumping good read'
Observer

'Sykes's wonderfully clear book should be compulsory reading for politicians . . . [he] provides an eye-opening guide to the new branch of science that is changing the human race's view of itself'
Literary Review

'A great scientific adventure'
Eastern Daily Press

'Sykes writes from imaginative sympathy with the harsh, short lives of these women'
Daily Telegraph

'Sykes's scientific discoveries are fascinating in their own right. But where he really scores is in the way he enables the lives of our ancestors to be seen afresh, and to cast new light on our own existence'
Yorkshire Evening Post

'A very readable introduction to some significant scientific advances'
Independent

'A signal success, effortlessly bringing the reader up to its breezy speed, with fascinating case histories ranging from the remains of the Russian royal family to pet golden hamsters'
Evening Standard

'A book that is moving and inspiring and a world away from dry science'
Publishing News

Also by Bryan Sykes

ADAM'S CURSE

and published by Corgi Books

BRYAN SYKES

— THE —

SEVEN DAUGHTERS OF EVE

CORGI BOOKS

THE SEVEN DAUGHTERS OF EVE
A CORGI BOOK : 0 552 15218 8

Originally published in Great Britain by Bantam Press,
a division of Transworld Publishers

PRINTING HISTORY
Bantam Press edition published 2001
Corgi edition published 2002

5 7 9 10 8 6

Typeset in Granjon by
Kestrel Data, Exeter, Devon.

Corgi Books are published by Transworld Publishers,
61–63 Uxbridge Road, London W5 5SA,
a division of The Random House Group Ltd,
in Australia by Random House Australia (Pty) Ltd,
20 Alfred Street, Milsons Point, Sydney, NSW 2061, Australia,
in New Zealand by Random House New Zealand Ltd,
18 Poland Road, Glenfield, Auckland 10, New Zealand
and in South Africa by Random House (Pty) Ltd,
Endulini, 5a Jubilee Road, Parktown 2193, South Africa.

Printed and bound in Great Britain by
Cox & Wyman Ltd, Reading, Berkshire.

Papers used by Transworld Publishers are natural, recyclable
products made from wood grown in sustainable forests.
The manufacturing processes conform to the environmental
regulations of the country of origin.

To my mother

CONTENTS

CONTENTS

ACKNOWLEDGEMENTS

This book owes many things to many people. Do not imagine for a moment that everything reported here as coming from my laboratory is exclusively my own work. Modern science relies on teamwork and I have been fortunate to have had some very talented people in my research group over the years. In different ways they have all helped in creating this story. In particular I want to thank Martin Richards, Vincent Macaulay, Kate Bendall, Kate Smalley, Jill Bailey, Isabelle Coulson, Eileen Hickey, Emilce Vega, Catherine Irven, Linda Ferguson, Andrew Lieboff, Jacob Low-Beer and Chris Tomkins. In Oxford I must also thank Robert Hedges from the Radiocarbon Accelerator Unit for getting me started on all this, William James, Fellow of most Oxford colleges in his time, for his inspired suggestions along the way and, in London, Chris Stringer of the Natural History Museum for allowing me to drill holes into the fossils in his care. I am very grateful to Clive Gamble for his tutorials on the ancient world. I must also pay particular thanks to Professor Sir David Weatherall for not only tolerating but actually encouraging the performance of such exotic and seemingly pointless research in his Institute of Molecular Medicine in Oxford.

You may also gain the impression that my research group is the only one in the world doing this sort of work. It certainly is not and none of what I describe would have been possible without the pioneering work of, among many others, Luca Cavalli-Sforza, Alberto Piazza, Walter Bodmer, the late Allan Wilson, Svante Paabo, Mark Stoneking, Rebecca Cann, Douglas Wallace, Antonio Torroni, Mark Jobling and Peter Underhill. As you will see, we do not all necessarily agree with one another all of the time; but without them, and many others like them, mine would have been a much harder, and far duller journey.

Four people in particular have helped to bring this story into print. The quiet professionalism of my editor, Sally Gaminara, and the infectious enthusiasm of my agent, Luigi Bonomi, have kept me going. Add to that the thoroughness of Gillian Bromley, my copy editor, and the patience of Julie Sheppard, who typed up my scribbled notes, and few authors could have had more assistance.

I am indebted to the thousands of volunteers who, by giving me their DNA samples, have allowed me to peer into the secrets of their genetic past. Without them there would be no story to tell. Some names have been changed to protect anonymity. I particularly want to thank the government and people of Rarotonga in the Cook Islands for being exceptionally helpful, and Malcolm Laxton-Blinkhorn for his outstanding hospitality during my stays on this delightful island. And lastly, thanks to Janis, Jay, Sue and my son Richard, though only an embryo at the time, for coming with me.

B.S.

THE SEVEN
DAUGHTERS OF EVE

Legend:
- Ice sheet at maximum extent
- Territories of the Seven Daughters
- Coastline at the lowest point during the last Ice Age c.18-20,000 years ago

800 KILOMETRES
500 MILES

HELINA
VELDA
SEINE
RHINE
ALPS
PO
APEN
PYRENEES
EBRO
TAGUS
TARA
EL

The Seven Gardens of Eden: the lands of our ancestors

URALS

VOLGA

DNIEPER

URAL

DON

XENIA

KATRINE

CARPATHIANS

DANUBE

URSULA

TIGRIS

JASMINE

EUPHRATES

PROLOGUE

Where do I come from?

How often have you asked yourself that question? We may know our parents, even our grandparents; not far beyond that, for most of us the trail begins to disappear into the mist. But each of us carries a message from our ancestors in every cell of our body. It is in our DNA, the genetic material that is handed down from generation to generation. Within the DNA is written not only our histories as individuals but the whole history of the human race. With the aid of recent advances in genetic technology, this history is now being revealed. We are at last able to begin to decipher the messages from the past. Our DNA does not fade like an ancient parchment; it does not rust in the ground like the sword of a warrior long dead. It is not eroded by wind or rain, nor reduced to ruin by fire and earthquake. It is the traveller from an antique land who lives within us all.

This book is about the history of the world as revealed by genetics. It shows how the history of our species, *Homo sapiens*, is recorded in the genes that trace our ancestry back into the deep past, way beyond the reach of written records or stone inscriptions. These genes tell a story which begins over a hundred thousand years ago and whose latest chapters are hidden within the cells of every one of us.

It is also my own story. As a practising scientist, I am very lucky to have been around at the right time and able to take an active part in this wonderful journey into the past that modern genetics now permits. I have found DNA in skeletons thousands of years old and seen exactly the same genes in my own friends. And I have discovered that, to my astonishment, we are all connected through our mothers to only a handful of women living tens of thousands of years ago.

In the pages that follow, I will take you through the excitement and the frustrations of the front-line research that lies behind these discoveries. Here you will see what really happens in a genetics laboratory. Like any walk of life, science has its ups and downs, its heroes and its villains.

1

ICEMAN'S RELATIVE FOUND
IN DORSET

On Thursday 19 September 1991 Erika and Helmut Simon, two experienced climbers from Nuremberg in Germany, were nearing the end of their walking holiday in the Italian Alps. The previous night they had made an unscheduled stop in a mountain hut, planning to walk down to their car the next morning. But it was such a brilliantly sunny day that they decided instead to spend the morning climbing the 3,516 metre Finailspitze. On their way back down to the hut to pick up their rucksacks they strayed from the marked path into a gully partly filled with melting ice. Sticking out of the ice was the naked body of a man.

Though macabre, such finds are not so unusual in the high Alps, and the Simons assumed that this was the body of a mountaineer who had fallen into a crevasse perhaps ten or twenty years previously. The following day the site was revisited by two other climbers, who were puzzled by the old-fashioned design of the ice-pick that was lying nearby. Judging by the equipment, this alpine accident went back a good many years. The police were contacted and, after checking the records of missing climbers, their first thought was that the body was probably that of Carlo Capsoni, a music professor from Verona who had disappeared in the area in 1941. Only days later did it begin to dawn on everybody that this was not a modern death at all. The tool found beside the body was nothing like a modern ice-pick. It was much more like a prehistoric axe. Also nearby was a container made from the bark of a birch tree. Slowly the realization sank in that this body was not tens or even hundreds but thousands of years old. This was now an archaeological find of international importance.

The withered and desiccated remains of the Iceman, as he soon came to be known, were taken to the Institute of Forensic Medicine in Innsbruck, Austria, where he was stored, frozen, while an international team of scientists was assembled to carry out a minute examination of this unique find. Since my research team in Oxford had been the first in the world to recover traces of DNA from ancient human bones, I was called in to see whether we could find any DNA in the Iceman. It was the irresistible opportunity to become involved in such thrilling discoveries that had persuaded me to veer away from my career as a regular medical geneticist into this completely new field of science, which some of my colleagues regarded as a bizarre and eccentric diversion of no conceivable use or consequence.

By now, carbon-dating – measuring the decay of minute traces of naturally radioactive carbon atoms within the remains – had confirmed the great antiquity of the Iceman, placing him between 5,000 and 5,350 years old. Even though this was much older than any human remains I had worked with before, I was very optimistic that there was a good chance of success, because the body had been deep frozen in ice away from the destructive forces of water and oxygen which, slowly but surely, destroy DNA. The material we had to work with had been put in a small screw-capped jar of the sort used for pathology specimens. It looked awfully unremarkable: a sort of grey mush. When Martin Richards, my research assistant at the time, and I opened the jar and started to pick through the contents with a pair of forceps, it seemed to be a mixture of skin and fragments of bone. Still, though it might not have been much to look at, there was no obvious sign that it had begun to decompose, and so we set to work with enthusiasm and optimism. Sure enough, back in the lab in Oxford, when we put the small fragments of bone through the extraction process that had succeeded with other ancient samples, we did find DNA, and plenty of it.

In due course we published our findings in *Science*, the leading US scientific journal. To be perfectly honest, the most remarkable thing about our results was not that we had got DNA out of the body – by then this was a routine process – but that we had got exactly the same DNA sequence from the Iceman as an independent team from Munich. We had both shown that the DNA was clearly European by finding precisely the same sequence in DNA samples taken from living Europeans. You might think this was not much of a surprise, but there was a real possibility that the whole episode could have been a gigantic hoax, with a South American mummy helicoptered in and planted in the ice. The cold and intensely dry air of the Atacama desert of southern Peru and northern Chile has preserved hundreds of complete bodies buried in shallow graves, and it would not have been hard for a determined hoaxer to get hold of one of them. The much damper conditions in Europe reduce a corpse to a skeleton very quickly, so if this was a hoax the body had to have come from somewhere else, probably South America. It may sound far-fetched; but elaborate tricks have been played before. Remember Piltdown Man. This infamous fossil had been 'discovered' in a gravel pit in Sussex, England, in 1912. It had an ape-like lower jaw attached to a much more human-looking skull, and was heralded as the long sought-after 'missing link' between humans and the great apes – gorillas, chimpanzees and orang-utans. Only in 1953 was it revealed to be a hoax, when radiocarbon analysis, the same technique that was later used to date the Iceman, proved beyond any doubt that the Piltdown skull was modern. The perpetrator, who has never been identified, had combined the lower jaw of an orang-utan with a human braincase and chemically stained them both to look much older than they really were. The long shadow cast by the Piltdown Man fraud lingers even to this day, so the idea that the Iceman might have been a hoax was very much at the front of everyone's mind.

There were a number of press enquiries following the publication of our scientific article about the Iceman, and I found myself explaining how we had proved his European credentials. Had it been a hoax, the DNA would have shown it. The closest matches would have been with South Americans and not with Europeans. But it was Lois Rogers from the *Sunday Times* who asked the crucial question.

'You say you found exactly the same DNA in modern Europeans. Well, who are they?' she enquired in a tone which told me she expected me to know the answer straight away.

'What do you mean, who are they? They are from our collection of DNA samples from all over Europe.'

'Yes, but who?' persisted Lois.

'I have no idea. We keep the identities of the donors on a separate file, and anyway, samples are always given on the basis of a strict undertaking of confidentiality.'

After Lois rang off, I switched on my computer just to see which samples matched up with the Iceman. LAB 2803 was one of them, and the series prefix 'LAB' meant it was either from someone working in the laboratory or from a visitor or friend. When I checked the number against the database containing the names of the volunteers, I could scarcely believe my luck. LAB 2803 was Marie Moseley, and LAB 2803 had exactly the same DNA as the Iceman. This could only mean one thing. Marie was a relative of the Iceman himself. For reasons which I shall explain in detail in later chapters, there had to be an unbroken genetic link between Marie and the Iceman's mother, stretching back over five thousand years and faithfully recorded in the DNA.

Marie is an Irish friend, a management consultant from just outside Bournemouth in Dorset in southern England. Though not a scientist herself, she has an insatiable curiosity about genetics and had donated a couple of strands of her long red hair in the cause of science two years earlier. She is articulate, outgoing and very witty, and I was sure she could handle any publicity. When I rang to ask if she would mind if I gave her name to the *Sunday Times* she agreed at once, and the next edition carried a piece on her under the headline 'Iceman's relative found in Dorset'.

For a few weeks after that, Marie became an international celebrity. Of all the headlines that followed, I liked the one from the *Irish Times* best of all. Their reporter had asked Marie if she had been left anything by her celebrated predecessor. Shockingly, she revealed that she had not; so the story appeared as 'Iceman leaves one of our own destitute in Bournemouth'.

One of the strangest and, at first, surprising things about this story, and the reason I tell it here, is that Marie began to feel something for the Iceman. She had seen pictures of him being shunted around from glacier to freezer to post-mortem room, poked and prodded, opened up, bits cut off. To her, he was no longer just the anonymous curiosity whose picture had appeared in the papers and on television. She had started to think of him as a real person and as a relative – which is exactly what he was.

I became fascinated by the sense of connection that Marie had felt between herself and the Iceman. It began to dawn on me that if Marie could be genetically linked to someone long dead, thousands of years before any records were kept, then so could everyone else. Perhaps we only needed to look around us, at people alive today, to unravel the mysteries of the past. Most of my archaeologist friends found this proposition completely foreign to them. They had been brought up to believe that one could understand the past only by studying the past; modern people were of no interest. Yet I was sure that if DNA was inherited intact for hundreds of generations over thousands of years, as I had shown by connecting Marie and the Iceman, then individuals alive today were as reliable a witness to past events as any bronze dagger or fragment of pottery.

It seemed to me absolutely essential to widen my research to cover modern people. Only when much more was known about the DNA of living people could I hope to put the results from human fossils into any sort of context. So I set out to discover as much as possible about the DNA in present-day Europeans and people from many other parts of the world, knowing that whatever I found would have been delivered to us direct from their ancestors. The past is within us all.

My research over the intervening decade has shown that almost everyone living in Europe can trace an unbroken genetic link, of the same kind that connects Marie to the Iceman, way back into the remote past, to one of only seven women. These seven women are the direct maternal ancestors of virtually all 650 million

modern Europeans. As soon as I gave them names –
Ursula, Xenia, Helena, Velda, Tara, Katrine and
Jasmine – they suddenly came to life. This book tells
how I came to such an incredible conclusion and what is
known about the lives of these seven women.

I know that I am a descendant of Tara, and I want to
know about her and her life. I feel I have something in
common with her, more so than I do with the others.
By ways which I will explain, I was able to estimate
how long ago, and approximately where, all seven
women had lived. I reckoned that Tara lived in
northern Italy about 17,000 years ago. Europe was
in the grip of the last Ice Age, and the only parts of the
continent where human life was possible were in the far
south. Then, the Tuscan hills were a very different
place. No vines grew; no bougainvillaea decorated the
farmhouses. The hillsides were thickly forested with
pine and birch. The streams held small trout and
crayfish, which helped Tara to raise her family and
held the pangs of hunger at bay when the menfolk
failed to kill a deer or wild boar. As the Ice Age
loosened its grip, Tara's children moved round the
coast into France and joined the great band of hunters
who followed the big game across the tundra that was
northern Europe. Eventually, Tara's children walked
across the dry land that was to become the English
Channel and moved right across to Ireland, from whose
ancient Celtic kingdom the clan of Tara takes its name.

Soon after the conclusions of my research were
published, news of these seven ancestral mothers began
to appear in newspapers and on television all round the

world. Writers and picture editors used their imagination in finding contemporary analogues: Brigitte Bardot became the reincarnation of Helena; Maria Callas was Ursula; the model Yasmin le Bon was linked, naturally, with Jasmine; Jennifer Lopez became Velda. So many people rang us to find out which one they were related to that we had to set up a website to handle the hundreds of enquiries. We had stumbled across something very fundamental; something we were only just beginning to understand.

This book tells the story behind these discoveries and their implications for us all, not just in Europe but all over the world. It is a story of our common heritage and our shared forebears. It takes us from the Balkans in the First World War to the far islands of the South Pacific. It takes us from the present time back to the beginnings of agriculture and beyond, to our ancestors who hunted with the Neanderthals. Amazingly, we all carry this history in our genes, patterns of DNA that have come down to us virtually unchanged from our distant ancestors – ancestors who are no longer just an abstract entity but real people who lived in conditions very different from those we enjoy today, who survived them and brought up their children. Our genes were there. They have come down to us over the millennia. They have travelled over land and sea, through mountain and forest. All of us, from the most powerful to the weakest, from the fabulously wealthy to the miserably poor, carry in our cells the survivors of these fantastic journeys – our genes. We should be very proud of them.

My part in this story begins at the Institute of

Molecular Medicine in Oxford, where I am a professor of genetics. The Institute is part of Oxford University, though geographically and temperamentally removed from the arcane world of the college cloisters. It is full of doctors and scientists who are working away applying the new technologies of genetics and molecular biology to the field of medicine. There are immunologists trying to make a vaccine against AIDS, oncologists working out how to kill tumours by cutting off their blood supply, haematologists striving to cure the inherited anaemias which disable or kill millions each year in the developing world, microbiologists unravelling the secrets of meningitis and many others. It is an exciting place to work. I am based at the Institute because I used to work on inherited diseases of the skeleton, in particular on a horrible condition called *osteogenesis imperfecta*, better known as brittle bone disease. Babies born with the most severe form of this disease sometimes have bones so weak that when they take their first breath, all the ribs fracture and they suffocate and die. We were researching the cause of this tragic disease and had traced it to tiny changes in the genes for collagen. Collagen is the most important and abundant protein in bones and it supports them in much the same way as steel rods strengthen reinforced concrete. It made sense that if collagen failed because of a fault in the gene, the bones would break. The research involved finding out a lot about the way collagen and its genes varied in the general population – and it was through this work that, in 1986, I came to meet Robert Hedges.

Robert runs the carbon-dating laboratory for archaeological samples in Oxford. He had been thinking about ways of getting more information from the bones that passed through his lab, aside from just dating them by the radiocarbon method. Collagen is the main protein not only in living bones but also in dead ones, and it is the carbon in the surviving collagen that is used to date them. Robert wondered if there was any genetic information in these surviving fragments of ancient collagen, so he and I put together a research proposal to study them. Collagen, being a protein, is made of units called amino-acids, arranged in a particular sequence. As we shall see in the next chapter, the sequence of amino-acids in collagen, and all other proteins for that matter, is dictated by the DNA sequence of their genes. We hoped to discover the DNA sequence of the ancient collagen genes indirectly by determining the order of amino-acids in the fragments of protein that survived in Robert's old bones. We advertised for research assistants several times but got no response at all. We would have expected a flood of applications for a regular genetics post, and put this zero interest down to the unusual nature of the project. Disappointingly few scientists want to venture from the mainstream field of research at an early stage of their careers. For us, this lack of a recruit meant we had to put back the start of the project by a year. Although very frustrating at the time, the delay proved to be a blessing in disguise – because, before the project got going, news came in of a new invention. A US scientist in California called Kary Mullis had dreamed up a way of amplifying tiny amounts of DNA – under perfect conditions, as little as a single molecule – in a test tube.

One warm Friday night in 1983 Mullis was driving along Highway 101 by the ocean; according to his account of events, 'the night was saturated with moisture and the scent of flowering buckeye'. As he drove, he was talking to his girlfriend, seated beside him, about some of the ideas he had been pondering to do with his work at a local biotech company. Like everyone else in the genetic engineering business, he was making copies of DNA in test tubes. This was a slow process because the molecules had to be copied one at a time. DNA is like a long piece of string, and the copying started at one end and finished at the other. Then it started at the beginning again and you got another copy. He was talking out loud about this and suddenly realized that if, instead of starting the copying at one end only, you started at *both* ends you would start what would effectively be a sustainable chain reaction. You would no longer just be making copies of the original but copies of copies, doubling the number at every cycle. Now, instead of two copies after two cycles and three copies after three cycles, you would double up after each cycle, producing two, four, eight, sixteen, thirty-two, sixty-four copies in six cycles instead of one, two, three, four, five and six. After twenty cycles you would have not just twenty copies but a million. It was a real 'Eureka' moment. He turned to his girlfriend to get her reaction. She had fallen asleep.

This invention, for which Kary Mullis rightly won the Nobel Prize for Chemistry in 1993, genuinely revolutionized the practice of genetics. It meant that you could now get an unlimited amount of DNA to work on from even the tiniest piece of tissue. A single hair or even a single cell was now all that was needed to produce as much DNA as you could ever want. The impact of Mullis's brainwave on our bone project was simply that I decided to forget about working on the collagen protein, which would have been horrendously difficult, and use the newly invented chain reaction to amplify what, if anything, was left of the DNA in the ancient bones. If it worked, then we would get vastly more information from the DNA than we would ever have got from the collagen. We would be going directly for the DNA sequence itself, rather than inferring it from the amino-acids. Much more importantly, we would be able to study *any* gene, not just the ones that controlled collagen.

At last we got an answer to our advertisement for a research assistant, and Erika Hagelberg joined the team. We were obviously not going to get anyone with previous experience in working with ancient DNA, because it had never been done before, but Erika's degree in biochemistry, combined with research posts in homoeopathy and in the history of medicine, reflected a combination of a solid scientific training and the catholic interests which suited the project. Besides, she was the only applicant. Now we needed some very old bones.

News came in during 1988 of an excavation going on in Abingdon, a few miles south of Oxford. A new supermarket was going up and the mechanical diggers had ploughed into a medieval cemetery. The local archaeology service had been given two months to excavate the site before the developers moved back in, so when Erika and I arrived, it was buzzing with activity. It was a hot and brilliantly sunny day and dozens of field assistants, stripped down to the bare essentials, were dotted all round the site scraping at the earth with trowels, rummaging around in deep pits or wading through water-filled trenches. Several skeletons lay half-exposed, encrusted with orange-brown earth, criss-crossed by strings which marked out a reference grid. As we gazed down at them, our prospects didn't look at all promising. Having worked with DNA for several years, I was trained to treat it with respect. DNA samples were always stored frozen at 70° below zero, and whenever you took DNA out of the freezer you were taught always to keep it in an ice bucket. If you forgot about it and the ice thawed then you had to throw the DNA out because, so everyone assumed, it would have degraded and been destroyed. No-one imagined it would last for more than a few minutes on the laboratory bench at room temperature, let alone buried underground for hundreds or even thousands of years.

Nevertheless, it was worth a try. We were allowed to take three thigh bones from the excavation away with us. Back in the lab we had to make two decisions: how to get the DNA out, and what section of DNA to choose for the amplification reaction. The first was easy enough. We knew that if there were any DNA left at all it would probably be bound up with a bone mineral

30

called hydroxyapatite. This form of calcium had been used in the past to absorb DNA during the purification process, so it seemed quite likely that the DNA would be stuck to the hydroxyapatite in the old bones. If that was the case, we had to think of a way of disengaging the DNA from the calcium.

We cut out small segments of bone with a hacksaw, froze them in liquid nitrogen, smashed them up into a powder, then soaked the powder in a chemical which slowly took out the calcium over several days. Fortunately, when all the calcium had been removed, there was still something left at the bottom of the tube – a sort of grey sludge. We guessed this was the remnants of the collagen and other proteins, bits of cells, maybe some fat – and, we hoped, a few molecules of DNA. We decided to get rid of the protein using an enzyme. Enzymes are the catalysts of biology, making things happen much more quickly than they otherwise would. We chose an enzyme which digests protein, rather like the ones in a biological washing powder which get rid of blood and other stains for the same reason. Then we got rid of the fat with chloroform. We cleaned what was left with phenol, a revolting liquid which is the base for carbolic soap. Even though phenol and chloroform are both brutal chemicals, we knew they did not harm DNA. What remained was a teaspoonful of pale brown fluid which, theoretically at least, should contain the DNA – if there was any. There would be at best only a few molecules, so we had to use the new DNA amplification reaction to boost the yield before we could carry out the next steps.

The essence of the amplification reaction is to adapt the system for copying DNA that cells use. Into the tube go the raw materials for DNA construction. First in is another enzyme, this time one used for copying DNA; it is called a polymerase and gives the reaction its scientific name – the *polymerase chain reaction* or PCR for short. Next, a couple of short DNA fragments are added to direct the polymerase enzyme to the segment of the original DNA that is to be amplified and ignore everything else. Finally, the raw materials – the nucleotide bases – for building new DNA molecules go into the mix along with a few ingredients, like magnesium, to help things along. Plus, of course, the stuff you want to amplify – in our case, an extract of the Abingdon bone containing, we hoped, a few molecules of very old DNA.

Then we had to decide which gene to amplify. Because we knew there wasn't going to be much, if any, DNA left in the bone extract we decided to maximize our chances by choosing something called mitochondrial DNA. We chose mitochondrial DNA for the simple reason that cells have upwards of a hundred times more of it than any other gene. As we will see, mitochondrial DNA turns out to have special properties which make it absolutely ideal for reconstructing the past; but in the first instance, we chose it as our target simply because there was so much more of it than any other type of DNA. If there was any DNA at all left in the Abingdon bones, then our best chance of finding it was by targeting mitochondrial DNA.

So, into the reaction went all the ingredients

necessary for amplifying mitochondrial DNA, plus a few drops of the precious bone extract. To get the reaction to fire in the tube you need to boil it, cool it, warm it up for a couple of minutes; then boil it again, cool it, warm it up . . . and go on repeating this cycle at least twenty times. Modern genetics laboratories are full of machines for doing this reaction automatically. But not then. Back in the 1980s the only machine on the market cost a fortune, and there was no money for one in our budget. The only way to do the reaction was to sit with a stop-watch in front of three water baths, one boiling, one cold and one warm, and move the test tube by hand from one bath to the next every three minutes. Then do it again. And again. For three and a half hours. I only tried it once. The reaction didn't work and I was bored stiff. There had to be a better way. What about using an electric kettle? I spent the next three weeks with wires, timers, thermostats, relays, copper tubing, a washing-machine valve and my kettle from home. In the end I had a device that did all the right things. It boiled. It cooled (very fast) when the washing-machine valve opened and let cold tap-water into the coils of copper tubing. And it warmed up. And it worked.

We could see that the machine (christened the 'Genesmaid', after the tea-making device people of a certain age regard as an essential bedroom accessory) had managed to get the amplification reaction to work not only with a control experiment using modern DNA but also, very faintly, with the Abingdon bone extract. By comparing its sequence to those published in scientific papers, it didn't take us long to prove that the DNA was genuinely human. We had done it. Here, in front of our very eyes, was the DNA of someone who had died hundreds of years ago. It was DNA resurrected, literally, from the grave.

Now, looking back, it is hard for me to believe that the research set in motion by the recovery of DNA from those crumbling bones in the Abingdon cemetery, the bones which looked so unpromising when I first saw them half-buried in the earth, should lead over the following years to such profound conclusions about the history and soul of our species. As my story unfolds you will see that, like most scientific research, this was not a seamless progression towards a well-defined goal. It was more like a series of short hops, each driven as much by opportunity, personal relationships, financial necessity and even physical injury as by any rational strategy. There was no set path towards the discovery of the Seven Daughters of Eve. The research just moved a little bit at a time, mostly forwards, towards the next dimly visible goal, informed by what had gone before but ignorant of what lay ahead.

At the time, though our result was a great triumph, strangely enough it didn't feel like it. I think Erika and I were too heavily involved in the details to appreciate the significance of what we had achieved. Besides, by then we were not getting on at all well. Tension had been building for weeks because, for some reason, Erika and I did not seem to be working together effectively. Only much later did I start to realize what our breakthrough could mean, not only for science but for popular history as well. That would come later; at the moment we had more pressing claims on our attention. I had heard on the grapevine that other research teams were also looking for DNA in old bones. This meant we had to get our work published with maximum speed, otherwise there was a real danger that we would be scooped. What counts in science is not being the first to do an experiment but being the first to publish the results. If someone else published even a day before we did, then they would claim the prize. Fortunately, the editor of the scientific journal *Nature* was persuaded to rush our paper into print in record time, and it was published just before Christmas 1989.

I was quite unprepared for what happened next. Although my previous research on brittle bone disease had occasionally been covered in the local papers and even once or twice in the nationals, it could not be said that any new result had sparked off a media frenzy. So it was a new experience when I got into work next day to find the phone constantly ringing with press enquiries. A few years previously I had actually spent three months in London as a reporter for ITN, which runs the television news service for the main commercial terrestrial channels in the UK. This venture was part of a well-intentioned fellowship scheme run by the Royal Society, designed to bridge the gap between science and the media. I was attracted to it by the generous expenses with which I hoped to pay off my bank overdraft. In fact, I ended up owing more money than I had to start with, not least because of the amount of time I spent in bars and restaurants with the well-heeled professionals. One night, for instance, I was precocious enough to offer to buy a drink for one well-known presenter. 'Thanks, dear boy, I'll have a bottle of Bollinger,' came the great man's answer. What could I do but comply? Still, though a financial disaster of major proportions, those few months taught me many things about the news media, including the way to trim my replies to reporters' questions down to the simple sentences I knew they wanted.

After a morning of fielding enquiries about our scientific paper, I was beginning to feel a little bored with explaining in one sentence what DNA was, etc. etc. By the time the science correspondent of the *Observer* rang, this ennui had got the better of me. Having gone through the standard questions, he asked what could be done now that DNA could be recovered from archaeological remains. I replied that one possibility was that we might be able to tell whether or not the Neanderthals had become extinct. A perfectly reasonable reply and, as it turned out, a correct forecast. Then I slipped in: 'Of course we will also be able to solve questions that have puzzled scholars for centuries – like whether Rameses II was a man or a woman.' As far as I know, not a single scholar has ever entertained this possibility for a second. No-one has ever had the slightest doubt that the great pharaoh was a man. And yet, on the following Sunday, underneath his likeness, I read the caption 'King/Queen Rameses II'.

Many years later I had the good fortune to be invited to the opening of the new Egyptology gallery in the British Museum in London. At dinner that evening in the magnificent Egyptian Sculpture Gallery, my place was set directly opposite the huge granite statue of Rameses. He was looking down right at me with his unnervingly benign and omniscient gaze. I knew at once that he had heard about my joke at his expense, and that I was going to be in big trouble in the afterlife.

One of the most difficult things about getting ancient DNA out of old bones is that, unless you are extremely careful, you end up amplifying modern DNA, including your own, instead of the fossil's. Even when it is present, the old DNA is pretty shattered. Chemical changes, mostly brought about by oxygen, slowly change the structure of the DNA so that it starts breaking down into smaller and smaller fragments. If even the tiniest speck of modern DNA gets into the reaction then the polymerase copying enzymes, which don't realize that you are trying to amplify the worn out little scraps of ancient DNA, concentrate their efforts on the pristine modern stuff and, not knowing any better, produce millions of copies of that instead. So it looks as though the reaction has been a great success. You put a drop of ancient bone extract in at the beginning and get masses of DNA out at the end. Only when you analyse it further do you realize that it's your own DNA, not that from the fossil at all.

Although we were fairly sure this hadn't happened with the Abingdon bone, we thought one way of checking would be by getting DNA from old animal rather than old human bones. It would then be very easy to tell whether we had amplified animal DNA – the real thing – or human DNA, which would have to be a contaminant. The best source of sufficiently old animal bones we could think of was the wreck of the *Mary Rose*. This magnificent galleon had sunk during an engagement with a French invasion fleet off Portsmouth in 1545. Very few of the crew survived. For over four hundred years the wreck lay in the mud under 14 metres of water until it was raised in 1982 and put on display in a museum in Portsmouth harbour, where it is still being drenched with a solution of water and anti-freeze to prevent its timbers from buckling. As well as the skeletons of the unfortunate crew, hundreds of animal and fish bones were recovered from the wreck. The ship had been full of supplies when it sank, and among these were sides of beef and pork and barrels of salted cod. We persuaded the museum curator to let us have a pig rib to try. Because it had spent most of its life (after death, that is) buried in the oxygen-free ooze at the bottom of the Solent, the rib was in very good condition and we managed to get lots of DNA from it without much trouble. We analysed it – and there was no doubt at all that it was from a pig and not a human.

The point of telling you all this is not to take you through our experiments one by one, but to explain the reaction when the result was published. More phone calls and more headlines – of which my favourite is from the *Independent on Sunday*: 'Pig brings home the bacon for DNA'. This was going to be fun.

2

SO, WHAT IS DNA AND
WHAT DOES IT DO?

All of us are aware, as people must have been for millennia, that children often resemble their parents and that the birth of a child follows nine months after sexual intercourse. The mechanism for inheritance remained a mystery until very recently, but that didn't stop people from coming up with all sorts of theories. There are plenty of references in classical Greek literature to family resemblances, and musing on the reasons for them was a familiar pastime for early philosophers. Aristotle, writing around 335 BC, speculated that the father provided the pattern for the unborn child and the mother's contribution was limited to sustaining it within the womb as well as after birth. This idea made perfect sense to the patriarchal attitudes of Western civilization at the time. It was only reasonable that the father, the provider of wealth and status, was also the architect of all his children's features and nature. This was not to underestimate the necessity of choosing a suitable wife. After all, seeds planted in a good soil always do better than those put into a poor one. However, there was a problem and it was one that was to haunt women for a long time to come.

If children are born with their father's design, how was it that men had daughters? Aristotle was challenged on this point during his lifetime, and his answer was that all babies would be the same as their fathers in every respect, including being male, unless they were somehow 'interfered with' in the womb. This 'interference' could be relatively minor, leading to such trivial variations as a child having red hair instead of black like his father; or it could be more substantial – leading to major ones such as being deformed or female. This attitude has had serious consequences for many women throughout history who have found themselves discarded and replaced because they failed to produce sons. This ancient theory developed into the notion of the *homunculus*, a tiny, preformed being that was inoculated into the woman during sexual intercourse. Even as late as the beginning of the eighteenth century the pioneer of microscopy, Anthony van Leewenhoek, imagined he could see tiny homunculi curled up in the heads of sperm.

Hippocrates, whose name is commemorated in the oath that newly qualified doctors used to take (some still do), had a less extreme view than Aristotle which did give women a role. He believed that both men and women produced a seminal fluid, and that the characteristics of the baby were decided by which parts of the fluid prevailed when they mixed after copulation. A child might have its father's eyes or its mother's nose as a result of this process; if neither parent's fluid prevailed for a particular characteristic, the child might be somewhere in between, having, for example, hair of a colour that was intermediate between the two parents.

This theory was much more obviously connected to most people's experience of real life. 'He's just like his father' or 'She's got her mother's smile' and other similar observations are repeated millions of times every day throughout the world. The idea that the parents' characteristics are somehow blended in the offspring was the predominant belief among scientists until the end of the nineteenth century. Darwin certainly knew no better, and it was one reason why he could never find a suitable mechanism to explain his theory of natural selection; for anything new and favourable would be continually diluted out by the blending process at each generation. Even though geneticists today scoff at such apparent ignorance among their predecessors, I wouldn't mind betting that a theory of blending is, even now, a perfectly satisfactory explanation for what most people observe with their own eyes.

Eventually, two practical developments in the nineteenth century provided key clues to what was really going on. One was the invention of new chemical dyes for the textile industry, and the other was a change in the way microscope lenses were ground which made big improvements in their performance. Greater magnification meant that individual cells were now easily visible; and their internal structure was revealed when they were stained with the new dyes. Now the process of fertilization, the fusion of a single large egg cell and a single small, determined sperm, could be observed. When cells divided, strange thread-like structures could be seen assembling and then separating equally into the two new cells. Because they stained very brightly with the new dyes these curious structures became known as *chromosomes* – from Greek, meaning literally, 'coloured bodies' – years before anyone had a clue about what they did.

During fertilization, one set of these strange threads seemed to come from the father's sperm and another set from the mother's egg. This was just what had been predicted by the man universally acknowledged as the father of genetics, Gregor Mendel, a monk in the town of Brno in the Czech republic who laid the foundation for the whole of genetics from his experimental breeding of peas in the monastery garden in the 1860s. He concluded that whatever it was that determined heredity would be passed on equally from both parents to their offspring. Unfortunately he died before he ever saw a chromosome; but he was right. With the important exception of mitochondrial DNA (of which we shall have much more to say later) and the chromosomes that determine sex, genes – specific pieces of genetic coding that occur in the chromosomes – are inherited equally from both sets of parents. The essential part played by chromosomes in heredity and the fact that they must contain within them the secrets of inheritance was already well established by 1903. But it took another fifty years to discover what chromosomes are made of and how they worked as the physical messengers of heredity.

In 1953 two young scientists working in Cambridge, James D. Watson and Francis Crick, solved the molecular structure of a substance which had been known about for a long time and largely thought of as dull and unimportant. As if to emphasize its obscurity, it was given a really long name, *deoxyribonucleic acid*, now happily abbreviated to DNA. Although a few experiments had implicated DNA in the mechanism of inheritance, the smart money was on proteins as the hereditary material. They were complicated, sophisticated, had twenty different components (the amino-acids) and could assume millions of different forms. Surely, the thinking went, only something really complicated could manage such a monumental task as programming a single fertilized egg cell to grow into a fully formed and functional human being. It couldn't possibly be this DNA, which had only four components. Admittedly it was in the right place, in the cell nucleus; but it probably did something very dull like absorbing water, rather like bran.

Despite the general lack of interest in this substance shown by most of their scientific contemporaries, Watson and Crick felt sure it held the key to the chemical mechanism of heredity. They decided to have a crack at working out its molecular structure using a technique that was already being used to solve the structure of the more glamorous proteins. This entailed making long crystalline fibres of purified DNA and bombarding them with X-rays. As the X-rays entered the DNA, most went straight through and out the other side. But a few collided with the atoms in the molecular structure and bounced off to one side where they were detected by sheets of X-ray film – the same kind of film that hospital radiographers still use to get an image of a fractured bone. The deflected X-rays made a regular pattern of spots on the film, whose precise locations were then used to calculate the positions of atoms within the DNA.

After many weeks spent building different models with rods and sheets of cardboard and metal to represent the atoms within DNA, Watson and Crick suddenly found one which fitted exactly with the X-ray pattern. It was simple, yet at the same time utterly marvellous, and it had a structure that immediately suggested how it might work as the genetic material. As they put it with engaging self-confidence in the scientific paper that announced the discovery: 'It has not escaped our notice that the specific pairings we have postulated immediately suggest a possible copying mechanism for the genetic material.' They were absolutely right, and were rewarded by the Nobel Prize for Medicine and Physiology in 1962.

One of the essential requirements for the genetic material had to be that it could be faithfully copied time and again, so that when a cell divides, both of the two new cells – the 'daughter cells', as they are called – each receive an equal share of the chromosomes in the nucleus. Unless the genetic material in the chromosomes could be copied every time a cell divided it would very soon run out. And the copying had to be very high quality or the cells just wouldn't work. Watson and Crick had discovered that each molecule of

DNA is made up of two very long coils, like two intertwined spiral staircases – a 'double helix'. When the time comes for copies to be made, the two spiral staircases of the double helix disengage. DNA has just four key components, which are always known by the first letters of their chemical names: A for adenine, C for cytosine, G for guanine and T for thymine. Formally they are known as *nucleotide bases* – 'bases' for short. You can now forget the chemicals and just remember the four symbols 'A', 'C', 'G' and 'T'.

The breakthrough in solving the DNA structure came when Watson and Crick realized that the only way the two strands of the double helix could fit together properly was if every 'A' on one strand is interlocked with a 'T' directly opposite it on the other strand. Just like two jigsaw pieces, 'A' will fit perfectly with 'T' but not with 'G' or 'C' or with another 'A'. In exactly the same way, 'C' and 'G' on opposite strands can fit only with each other, not with 'A' or 'T'. This way *both* strands retain the complementary coded sequence information. For example, the sequence 'ATTCAG' on one strand has to be matched by the sequence 'TAAGTC' on the other. When the double helix unravels this section, the cell machinery constructs a new sequence *'TAAGTC'* opposite 'ATTCAG' on one of the old strands and builds up *'ATTCAG'* opposite 'TAAGTC' on the other. The result is two new double helices identical to the original. Two perfect copies every time. Preserved during all this copying is the sequence of the four chemical letters. And what is the sequence? It is information pure and simple. DNA doesn't actually do anything itself. It doesn't help you breathe or digest your food. It just instructs other things how to do it. The cellular middle managers which receive the instructions and do the work are, it turns out, the proteins. They might look sophisticated, and they are; but they operate under strict directions from the boardroom, the DNA itself.

Although the complexity of cells, tissues and whole organisms is breathtaking, the way in which the basic DNA instructions are written is astonishingly simple. Like more familiar instruction systems such as language, numbers or computer binary code, what matters is not so much the symbols themselves but the order in which they appear. Anagrams, for example 'derail' and 'redial', contain exactly the same letters but in a different order, and so the words they spell out have completely different meanings. Similarly, 476,021 and 104,762 are different numbers using the same symbols laid out differently. Likewise, 001010 and 100100 have very different meanings in binary code. In exactly the same way the order of the four chemical symbols in DNA embodies the message. 'ACGGTA' and 'GACAGT' are DNA anagrams that mean completely different things to a cell, just as 'derail' and 'redial' have different meanings for us.

So, how is the message written and how is it read? DNA is confined to the chromosomes, which never leave the cell nucleus. It is the proteins that do all the real work. They are the executives of the body. They are the enzymes which digest your food and run your metabolism; they are the hormones that coordinate what is happening in different parts of your body. They are the collagens of the skin and bone, and the haemoglobins of the blood. They are the antibodies that fight off infection. In other words, they do everything. Some are enormous molecules, some are tiny. What they all have in common is that they are made up of a string of sub-units, called amino-acids, whose precise order dictates their function. Amino-acids in one part of the string attract amino-acids from another part, and what was a nice linear string crumples up into a ball. But this is a ball with a very particular shape, that then allows the protein to do what it was made for: being a catalyst for biological reactions if it is an enzyme, making muscles if it is a muscle protein, trapping invading bacteria if it is an antibody, and so on. There are twenty amino-acids in all, some with vaguely familiar names like lysine or phenylalanine (one of the ingredients of the sweetener aspartame) and others most people haven't come across, like cysteine or tyrosine. The order in which these amino-acids appear in the protein precisely determines its final shape and function, so all that is required to make a protein is a set of DNA instructions which define this order. Somehow the coded information contained in the DNA within the cell nucleus must be relayed to the protein production lines in another part of the cell.

If you can spare one, pluck out a hair. The translucent blob on one end is the root or follicle. There are roughly a million cells in each hair follicle, and their only purpose in life is to make hair, which is mainly made up of the protein keratin. As you pulled the hair out, the cells were still working. Imagine yourself inside one of these cells. Each one is busy making keratin. But how do they know how to do it? The key to making any protein, including keratin, is just a matter of making sure that the amino-acids are put in the right order. What is the right order? Go and look it up in the DNA which is on the chromosomes in the cell nucleus. A hair cell, like every cell in the body, has a full set of DNA instructions, but you only want to know how to make keratin. Hair cells are not interested in how to make bone or blood, so all those sections of DNA are shut down. But the keratin instruction, the keratin *gene*, is open for consultation. It is simply the sequence of DNA symbols specifying the order of amino-acids in keratin.

The DNA sequence in the keratin gene begins like this: ATGACCTCCTTC . . . (etc. etc.). Because we are not used to reading this code it looks like a random arrangement of the four DNA symbols. However, while it might be unintelligible to us, it is not so to the hair cell. This is a small part of the code for making keratin, and it is very simple to translate. First the cell reads the code in groups of three symbols. Thus ATGACCTCCTTC becomes ATG–ACC–TCC–TTC. Each of these groups of three symbols, called a triplet, specifies a particular amino-acid. The first triplet ATG is the code for the amino-acid methionine, ACC stands for threonine, TCC for serine, TTC for phenylalanine and so on. This is the genetic code which is used by all genes in the cell nuclei of all species of plants and animals.

The cell makes a temporary copy of this code, as if it were photocopying a few pages of a book, then dispatches it to the protein-making machinery in another part of the cell. When it arrives here, the production plant swings into action. It reads the first triplet and decodes it as meaning the amino-acid methionine. It takes a molecule of methionine off the shelf. It reads the second triplet for the amino-acid threonine, takes a molecule of threonine down and joins it to the methionine. The third triplet means serine, so a molecule of serine gets tacked on to the threonine. The fourth triplet is for phenylalanine, so one of these is joined to the serine. Now we have the four amino-acids specified by the DNA sequence of the keratin gene assembled in the correct order: methionine–threonine–serine–phenylalanine. The next triplet is read, and the fifth amino-acid is added, and so on. This process of reading, decoding and adding amino-acids in the right order continues until the whole instructions have been read through to the end. The new keratin molecule is now complete. It is cut loose and goes to join hundreds of millions of others to form part of one of the hairs that are growing out of your scalp. Well, it would if you had not pulled it out.

3

FROM BLOOD GROUPS TO GENES

There are few things more distinctive about a person than their hair. It is one of the very first features we ask for in any description of a new baby, a stranger or a wanted criminal. Dark or blond, wavy or straight, thick or balding: all these different possibilities add immediately to the picture we build up in our minds of someone we have never met. We certainly know how to manipulate the way our own hair appears. Salons are full as we pay to have our hair cut and shaped. Pharmacy shelves are lined with products to lighten, darken, straighten and curl. We are all working to make the best of the hair we were born with; but it is our genes which deal out the basic raw material. The difference between a natural redhead and a blonde lies in a difference in their DNA. Within the genes for keratin and the many others involved in the process of growing hair are small differences in the DNA sequence. These are responsible for giving the hair different characteristics of colour and texture. Most of these genes have yet to be identified, but they are certain to be inherited from both parents, although not necessarily in a straightforward way – which is why it is a fairly frequent occurrence that a new baby does not have the hair colour of either of its parents.

Hair type is a highly visible distinguishing feature by which we tell individuals apart, but by far the greatest inherited differences between us are invisible and remain hidden unless something brings them to our attention. The first of these inherited differences to be revealed were the blood groups. You cannot tell just by looking at someone which blood group he or she belongs to. You can't even tell by simply looking at a drop of their blood. All blood looks pretty much the same. It is only when you begin to mix blood from two people that the differences begin to make themselves apparent; and, since no-one had any reason to mix one person's blood with another until blood transfusions were invented, our blood groups stayed hidden.

The first blood transfusions were recorded in Italy in 1628, but so many people died from the severe reactions that the practice was banned there, as well as in France and England. Though there were some experimental transfusions using lamb's blood, notably by the English physician Richard Lower in the 1660s, the results were no better and the idea was given up for a couple of centuries. Transfusions with human blood started up again in the middle of the nineteenth century, to combat the frequently fatal haemorrhages that occurred after childbirth, and by 1875 there had been 347 recorded transfusions. But many patients were still suffering the sometimes fatal consequences of a bad reaction to the transfused blood.

By that time, scientists were beginning to discover the differences in blood type that were causing the problem. The nature of the reaction of one blood type with another was discovered by the French physiologist Léonard Lalois when, in 1875, he mixed the blood of animals of different species. He noticed that the blood cells clumped together and frequently burst open. But it was 1900 before the biologist Karl Landsteiner worked out what was happening and discovered the first human blood group system, which divides people into Groups A, B, AB and O. When a donor's ABO blood group matches that of the patient receiving the transfusion, there is no bad reaction; but if there is a mismatch, the cells form clumps and break open, causing a severe reaction. There is some historical evidence that the Incas of South America had practised transfusions successfully. Since we now know that most native South Americans have the same blood group (Group O), the Inca transfusions would have been much less dangerous than attempts in Europe, because there was an excellent chance that both donor and patient would belong to Group O and thus be perfectly matched.

Unlike the complicated genetics which governs the inheritance of hair, which is still not fully understood, the rules for inheriting the ABO blood groups turned out to be very simple indeed. Precisely because the genetics were so straightforward and could be followed easily from parents to offspring, blood groups were widely used in cases of contested paternity until recently, when they were eclipsed by the much greater precision of genetic fingerprints. Their significance for our story in this book is that it was the blood groups which first launched genetics on to the world stage of human evolution. For this debut we need to go back to the First World War and to a scientific paper delivered to the Salonika Medical Society on 5 June 1918. It was translated and published the following year in the leading British medical periodical *The Lancet* under the title 'Serological differences between the blood of different races: the results of research on the Macedonian Front'. To give you a flavour of the sort of thing *The Lancet* published in those days, the article was sandwiched between a discourse by the eminent surgeon Sir John Bland-Sutton on the third eyelid of reptiles and a War Office announcement that those nurses who had been mentioned in dispatches for their work in Egypt and France would soon be getting a certificate from the King showing his appreciation.

The authors of the blood group paper were a husband and wife team, Ludwik and Hanka Herschfeld, who worked at the central blood group testing laboratory of the Royal Serbian Army, which was part of the Allied force fighting against the Germans. The First World War had a great influence on bringing blood transfusion practice towards its modern standards. Before the war it had been customary for physicians with a patient who needed a transfusion to test the blood groups of friends and relatives until a match was found, then bleed the donor and immediately give the blood to the patient. The high demand for transfusions on the battlefields of Europe meant that ways had to be found to store donated blood in blood banks ready for immediate use. All soldiers had their blood group tested and recorded so that, should they need an urgent transfusion to treat a serious battlefield wound, compatible blood of the correct type could be immediately drawn from the blood bank.

Ludwik Herschfeld had already demonstrated, some years earlier, that blood groups A and B followed the basic genetic rules laid out by Gregor Mendel. He was not sure what to make of blood group O and set it aside, though it was later shown to obey the same rules. Herschfeld saw the war as an opportunity to discover more about blood groups, and in particular how they compared in different parts of the world. The Allies drew soldiers from many different countries, and the Herschfelds set out to collate the blood group results from as many different nationalities as possible. It was a lot of work, but easier in wartime than later, when the research would, as they put it, 'have necessitated long years of travel'. For the obvious military reason that they were on the other side, they did not have the German data to hand, and the figures published in *The Lancet* were 'quoted from memory'.

When the Herschfelds came to review the results of their work, they found very big differences in the frequencies of blood groups A and B in soldiers who came from different 'races' as they called them. Among the Europeans, the proportions were around 15 per cent blood group B and 40 per cent blood group A. The proportion of men with blood group B was higher in troops drawn from Africa and Russia, reaching a peak of 50 per cent in regiments of the Indian Army fighting with the British. As the proportion of blood group B increased, there was a corresponding decrease in the frequency of blood group A.

In drawing their conclusions, the Herschfelds did not flinch from interpreting the significance of their results on a grand scale. They decided that humans were made up of two different 'biochemical races', each with its own origin: Race A, with blood group A, and Race B, with blood group B. Because Indians had the highest frequency of blood group B, they concluded that 'We should look to India as the cradle of one part of humanity.' As to how blood groups, and populations, spread, they go on: 'Both to Indo-China in the East and to the West a broad stream of Indians passed out, ever-lessening in its flow, which finally penetrated to Western Europe.' They were unsure about the origin of Race A and thought it might have come from somewhere around north or central Europe. We know now that their conclusions are complete nonsense; but they do illustrate that geneticists, then as now, are never shy of grandiose speculation.

The basic principle behind the evolutionary inferences drawn from the Herschfelds' blood group results was that 'races' or 'populations' that have similar proportions of the different blood groups are more likely to share a common history than those where the proportions are very different. This sounds like common sense, and it looks like a reasonable explanation for the similarities found in the different European armies. But there were also some surprises. For example, the blood group frequencies of soldiers from Madagascar and Russia were almost identical. Did this mean the Herschfelds had uncovered genetic evidence for a hitherto unrecorded Russian invasion of Madagascar, or even the reverse, an overwhelming Malagasy colonization of Russia? Or take the Senegalese from West Africa, who were almost as close in their blood group frequencies to the Russians as the English were to the Greeks, which seems a bit unusual to say the least. What was happening was that because they were working with just one genetic system – the only one available to them – their analysis produced what appear to be some very reasonable comparisons between populations and others that look distinctly odd.

In the years after the First World War, it fell to the American physician William Boyd to compile the abundant blood group data coming from transfusion centres throughout the world. As he did so, he saw inconsistencies of the Russia/Madagascar kind revealed by the original Herschfeld results time and again, so frequently, in fact, that he actively discouraged anthropologists from taking any notice of blood groups. Boyd quotes a letter from one frustrated correspondent: 'I tried to see what blood groups would tell me about ancient man and found the results very disappointing.' Even so, the unsuccessful attempts to explain human origins using blood groups had had their compensations for the liberal-minded Boyd. He wrote: 'In certain parts of the world an individual will be considered inferior if he has, for instance, a dark skin but in no part of the world does possession of a blood group A gene exclude him from the best society.'

After the Second World War, William Boyd's baton as compiler of blood group data from around the world passed to the Englishman Arthur Mourant. A native of Jersey in the Channel Islands, Mourant originally took a degree in geology but was unable to translate that training into a career. His very strict Methodist upbringing had caused him considerable emotional un-happiness, which he determined to resolve by becoming a psychoanalyst. To do this he decided first to study medicine and enrolled, at the relatively late age of thirty-four, in St Bartholomew's Medical School in London. This was in 1939, just before the outbreak of the Second World War. To avoid the German bombing raids on the capital, his medical school was moved from London to Cambridge, and it was here that he met R. A. Fisher, the most influential geneticist of his day. Fisher had been working out the genetics of the new blood groups which were being discovered, and he had become fascinated by the particularly convoluted inheritance of one of them – the Rhesus blood group. This new group had been discovered by Karl Landsteiner and his colleague Alexander Wiener in 1940 after they mixed human blood with the blood of rabbits that had themselves been injected with cells of the Rhesus monkey (hence the name). Fisher had come up with a complicated theory to account for the way in which the different sub-types within the group were passed down from parents to their children, and this was being violently attacked by Wiener who had offered a much simpler explanation. Imagine Fisher's delight when the new arrival, Arthur Mourant, discovered a large family of twelve siblings which provided the practical proof of his theory. Fisher found him a job at once, and the meticulous Mourant spent the rest of his working life compiling and interpreting the most detailed blood group frequency distribution maps ever produced. He never did become a psychoanalyst.

As well as being instrumental in getting Arthur Mourant a job, the Rhesus blood groups were also about to play a central role in what people were thinking about the origins of modern Europeans and in identifying the continent's most influential genetic population – the fiercely independent Basques of northwest Spain and south-west France. The Basques are unified by their common language, Euskara, which is unique in Europe in that it has no linguistic connection with any other living language. That it survives at all in the face of its modern rivals, Castilian Spanish and French, is remarkable enough. But two thousand years ago, it was only the disruption of imperial Roman administration in that part of the empire that saved Euskara from being completely swamped by Latin, which was the fate of the now extinct Iberian language in eastern Spain and south-east France. The Basques provided us with an invaluable clue to the genetic history of the whole of Europe, as we shall see later in the book, but their elevation to special genetic status only began when Arthur Mourant started to look closely at the Rhesus blood groups.

Most people have heard about the Rhesus blood

groups in connection with the medical condition known as 'haemolytic disease of the new-born' more commonly known simply as 'Rhesus baby syndrome'. This serious and often fatal condition affects the second or subsequent pregnancy of mothers who are 'Rhesus negative' – that is, who do not possess the Rhesus antigen on the surface of their red blood cells. What happens is this. When a Rhesus negative mother bears the child of a Rhesus positive father (whose red blood cells *do* carry the Rhesus antigen), there is a high probability that the foetus will be Rhesus positive. This is not a problem for the first child; but, when it is being born, a few of its red blood cells may get into the mother's circulation. The mother's immune system recognizes these cells, with their Rhesus antigen, as foreign, and begins to make antibodies against them. That isn't a problem for her until she becomes pregnant with her next child. If this foetus is also Rhesus positive then it will be attacked by her anti-Rhesus antibodies as they pass across the placenta. New-born babies affected in this way, who appear blue through lack of oxygen in their blood, could sometimes be rescued by a blood transfusion, but this was a risky procedure. Fortunately, 'blue baby syndrome' is no longer a severe clinical problem today. All Rhesus negative mothers are now given an injection of antibodies against Rhesus positive blood cells, so that if any do manage to get into her circulation during the birth of her first child they will be mopped up before her immune system has a chance to find them and start to make antibodies.

The significance of all this to the thinking about

European prehistory is that Mourant realized that having two Rhesus blood groups in a single population did not make any evolutionary sense. Even the simplest calculations showed that losing so many babies was not a stable arrangement. There was no problem if everybody had the same Rhesus type. It didn't matter whether this was Rhesus positive or Rhesus negative, just so long as it was all one or the other. It was only when there were people with different Rhesus types breeding together that these very serious problems arose. In the past, before blood transfusions and before the antibody treatment for Rhesus negative mothers, there must have been a lot of babies dying from haemolytic disease. This is a very heavy evolutionary burden, and the expected result of this unbalanced situation would be that one or other of the Rhesus blood groups would eventually disappear. And this is exactly what has happened – everywhere except in Europe. While the rest of the world is predominantly Rhesus positive, Europe stands out as having a very nearly equal frequency of both types. To Mourant, this was a signal that the population of Europe was a mixture that had not yet had time to settle down and eliminate one or other of the Rhesus types. His explanation was that modern Europe might be a relatively recent hybrid population of Rhesus positive arrivals from the Near East, probably the people who brought farming into Europe beginning about eight thousand years ago, and the descendants of an earlier Rhesus negative hunter-gathering people. But who were the Rhesus negatives?

Mourant came across the work of the French anthropologist H. V. Vallois, who described features of the skeletons of contemporary Basques as having more in common with fossil humans from about twenty thousand years ago than with modern people from other parts of Europe. Though this kind of comparison has since fallen into disrepute, it certainly catalysed Mourant's thinking. It was already known that Basques had by far the lowest frequency of blood group B of all the population groups in Europe. Could they be the ancient reservoir of Rhesus negative as well? In 1947 Mourant arranged to meet with two Basques who were in London attempting to form a provisional government and were keen to support any attempts to prove their genetic uniqueness. Like most Basques, they were supporters of the French Resistance and totally opposed to the fascist Franco regime in Spain. Both men provided blood samples and both were Rhesus negative. Through these contacts, Mourant typed a panel of French and Spanish Basques who turned out, as he had hoped, to have a very high frequency of Rhesus negatives, in fact the highest in the world. Mourant concluded from this that the Basques were descended from the original inhabitants of Europe, whereas all other Europeans were a mixture of originals and more recent arrivals, which he thought were the first farmers from the Near East.

From that moment, the Basques assumed the status of the population against which all ideas about European genetic prehistory were to be – and to a large extent still are – judged. The fact that they alone of all the west Europeans spoke a language which was unique in Europe, and did not belong to the Indo-European family which embraces all other languages of western Europe, only enhanced their special position.

The next leap forward came from the mathematical amalgamation of the vast amount of data that had accumulated from decades of research on individual systems like the different blood groups. This was accomplished by the man who has dominated the field for the past thirty years, Luigi Luca Cavalli-Sforza. We will meet him again later. Cavalli-Sforza, working with the Cambridge statistician Anthony Edwards, achieved this amalgamation using the earliest punched-card computing machines. By averaging across several genetic systems at once they managed to eliminate most of the bizarre and counter-intuitive conclusions that had discredited the anthropological applications of blood groups when they were worked on one at a time. The weakness of using just a single system was that two populations, like the Russians and the Malagasy, could end up with the same gene frequency just by chance rather than because of a common ancestry. This was far less likely to happen if several genes were compared, because the impact of a misleading result from one of them would be diluted out by the effect of the others. There were to be no more Russian invasions of Madagascar. None the less, the underlying principle remained the same. In an evolutionary sense, populations with similar gene frequencies were more likely to be closely related to each other than populations whose gene frequencies were very different.

Anthony Edwards explained his thinking in an ingenious article in *New Scientist* in 1965. He imagines a tribe that carries with it a pole along which are arrayed 100 discs which are either black or white. Every year, one disc, chosen at random, is changed to the other colour. When the tribe splits into two groups, each group takes with it a copy of the pole with the discs in their current order. The following year they each make one of the random changes to the discs. The next year they make another, the next year another and so on, continuing the custom of one random change every year. Since the changes they make are completely random, the order of the discs on the two poles becomes more and more dissimilar as each year passes. It follows that if you were to look at the poles carried by the two tribes you could estimate how long ago, in a relative sense, they separated from each other by the differences in the order of the black and white discs. Providing an absolute date was very difficult from the gene frequency data alone, but the comparative separation between the two tribes, known as the *genetic distance*, was a useful measure of their common ancestry. The bigger the genetic distance between them, the longer they had spent apart.

This was a clever image of the process of genetic change, called *genetic drift*, brought about by the random survival and extinction of genes as they pass from one generation to the next. This process leads to bigger and bigger differences in the frequencies of genes as time passes. Just like the order of discs in Edwards' analogy, gene frequencies can be used to

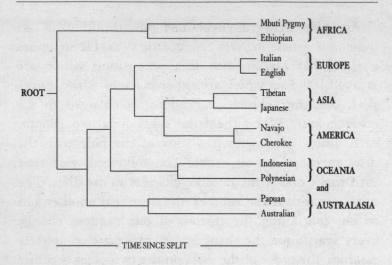

Figure 1

backtrack and work out how long ago two groups of people were once together as a single population. These groups could be villages, tribes or whole populations, and there is no limit to the number of groups that can be analysed in this way. If you do it for the whole world, the outcome is a diagram like Figure 1 above.

Along the right-hand side we have several 'populations' (I have picked two examples from each continent) and along the bottom is the genetic distance/time axis. This is what is called a population tree where the lines trace, from left to right, the estimated order in which 'populations' evolved and split from one another, as reconstructed from the assimilated frequencies of many different genes. At first glance, many of the groupings look quite sensible. The two European populations, the English and the Italians, are close together on two short 'branches' of the tree. The two native American tribes are connected together with their closest relatives in Asia, as we would expect if the first Americans crossed the Bering land bridge from Siberia to Alaska. The two populations from Africa are on a different branch from the rest of the world, which correctly emphasizes that continent's great antiquity as the cradle of human evolution. This is a much more sensible-looking tree than can be drawn from the First World War blood group data which, as well as allying Russia and Madagascar, entirely missed the importance of Africa. The reason for this, as noted earlier, is that the odd quirks that arose by chance with a single system, like the ABO blood groups, get ironed out by amalgamating the results from several different genes.

Edwards acknowledged that 'The resultant evolutionary trees will certainly not provide the last word on human evolution,' and offered the diagrams as a way of providing the genetic information in an understandable form. Unfortunately, the population trees first drawn with this admirable and modest intention were over-interpreted and became a source of contention. Among the several reasons for this is just the way they appear. They do look as if they are real evolutionary trees and have often been portrayed as exactly that. They could only be true evolutionary trees if human evolution really were a succession of population fissions along the lines of the splits that Edwards explains in his metaphor of the tribes with their poles and discs. Then and only then would the nodes, the points on the tree from which two lines diverge, represent a real entity. These would be the populations that existed before the splits, the proto-populations. But is that what really happened in human evolution? For instance, in the European part of the tree, was there ever such a thing as the proto-Anglo-Italian population which divided, never to meet again, and became the modern inhabitants of England and Italy? That might have been the case if the English and Italians became two different species as soon as they split and could never interbreed again. But they can, and they do, and they always have done. As we will discover later in the book, humans just did not evolve like this.

Perhaps the most serious objection to these trees is that their construction demands that the things at the end of the trees, the populations, be objectively defined. This process in itself segregates people into groups in ways that can tend to perpetuate racial classifications. It gives some sort of overall genetic number to something that does not really exist. There are certainly *people* who live in Japan and Tibet, but there is no genetic meaning to *the population* of Tibet or Japan, taken as a whole. As this book will show, objectively defined races simply do not exist. Even Arthur Mourant realized that fact nearly fifty years ago, when he wrote: 'Rather does a study of blood groups show a heterogeneity in the proudest nation and support the view that the races of the present day are but temporary integrations in the constant process of . . . mixing that marks the history of every living species.' The temptation to classify the human species into categories which have no objective basis is an inevitable but regrettable consequence of the gene frequency system when it is taken too far. For several years the study of human genetics got firmly bogged down in the intellectually pointless (and morally dangerous) morass of constructing ever more detailed classifications of human population groups.

Fortunately, there was a way out of this impasse. The break-out came with the publication of a scientific paper in *Nature* in January 1987 by the veteran US evolutionary biochemist, the late Allan Wilson, and two of his students, Rebecca Cann and Mark Stoneking, entitled 'Mitochondrial DNA and human evolution'. The centrepiece of this article was a diagram which bears a superficial resemblance to the trees I have just been criticizing. I have reproduced a small section of it here in Figure 2, with only sixteen individuals instead of the 134 in the original paper.

It is indeed an evolutionary tree; but this time the diagram means something. On the right of the tree the symbols at the tips of the branches represent not populations but the sixteen *individuals* that I have selected to illustrate the point, sixteen people from four different parts of the world: Africans, Asians, Europeans and Papuans from New Guinea. The first improvement over the other trees is that, unlike *populations*, there is no argument about whether *people* exist or not. They clearly do. The other improvement is that the nodes on the tree are also real people and not some hypothetical concept like a 'proto-population'. They represent the last common ancestors of the two

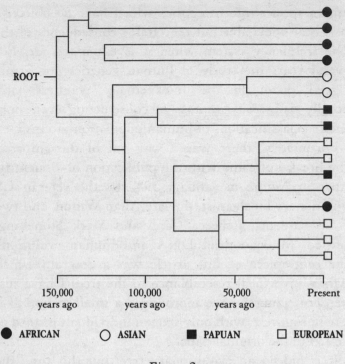

150,000
years ago   100,000
years ago   50,000
years ago   Present

● AFRICAN   ○ ASIAN   ■ PAPUAN   □ EUROPEAN

Figure 2

people who branch off from that point. The lines that connect the sixteen people on the diagram are drawn to reflect genetic differences between them in one very special gene called mitochondrial DNA whose unusual and useful properties I will introduce shortly. For reasons I shall explain in the next chapter, if two people have very similar mitochondrial DNA then they are more closely related, with respect to this gene, than two people with very different mitochondrial DNA. They have a common ancestor who lived more recently in the past, and so are joined by shorter branches on the diagram. People with very different mitochondrial DNA share a more remote common ancestor and are linked by longer branches.

To see how this works we can use again the metaphor of the tribe with its pole holding black and white discs. But this time the pole is the mitochondrial DNA and the tribe that split in two is a person who has two children. Both children inherit the same mitochondrial DNA, the genetic equivalent of the same pattern of discs on the pole. When they have their own children they pass on the mitochondrial DNA to them, and so it goes on down the generations. Very occasionally, random changes, called mutations, occur in the mitochondrial DNA which alter it a little bit at a time. These occur quite by chance when the DNA is being copied as cells divide. As time passes, more random changes are added to the DNA, which are then retained and passed on to future generations. Very slowly, the mitochondrial DNA of the descendants of that first individual, their common ancestor, becomes more and more different as more random mutations are introduced one at a time.

The lines on the tree in Figure 2 are reconstructions of the relationships among these sixteen people, worked out from the differences in their mitochondrial DNA, the exact nature of which we will examine shortly. But look for the moment at the tree itself. The deep trunk at the top has four Africans at the tips, while the other deep trunk contains individuals from the rest of the world *and* one more African. Within this 'rest of the world' trunk, close branches sometimes connect people from the same part of the world, like the Asians and Papuans at the top or the Europeans at the bottom. But they also sometimes connect individuals from different places, like the branch near the middle that links a Papuan with an Asian and two Europeans. What's going on? The deep split between the exclusively African 'trunk' and the rest of the world is another confirmation of the antiquity of Africa which the population trees also pick up. The confusion in the 'rest of the world' trunk is confirmation of exactly what Arthur Mourant had in mind. It is 'the mixing that marks the history of every living species'. Small wonder, then, that this diagram threw a very large spanner in the works of the population tree *aficionados*. It shows that genetically related individuals are cropping up all over the place, in all the wrong populations. You just cannot sustain the fundamental idea of a population being a separate biological and genetic unit if individuals within one population have their closest relatives within another.

Moreover, as we shall see in greater detail later on, by using the mutation process just described we can estimate the rate at which mitochondrial DNA changes with time. This means we can work out the timescales involved. When we do that, all the branches and the trunks converge to a single point, the 'root' of the tree, at about 150,000 years ago. This had to mean that the whole of the human species was much younger and more closely related than many people thought.

The impact of 'Mitochondrial DNA and human

evolution' was dramatic. It came down very firmly on one side of the argument about a fundamental question of human evolution. For many years there had been an intense and polarized debate on the origins of modern humans, based on different interpretations of fossil skeletons, mainly the skull. Both sides agreed that modern *Homo sapiens*, the species to which we all belong, originated in Africa. Both sides also agreed that an earlier type of human, called *Homo erectus*, was an evolutionary intermediate between ourselves and much older and more ape-like fossils. *Homo erectus* first appeared in Africa about two million years ago and by one million years ago, or perhaps even earlier, it had spread out to the warmer parts of the Old World. *Homo erectus* fossils have been found from Europe in the west to China and Indonesia in the east.

All that was — and is — agreed by both sides of the argument. What divides them is whether or not there was a much more recent spread of modern humans from Africa. The 'Out of Africa' school think there was, about 100,000 years ago, and that these new humans, our own *Homo sapiens*, completely replaced *Homo erectus* throughout its range. The opposing school of thought, the multi-regionalists, see clues in the fossils that suggest to them that *Homo sapiens* evolved directly from their local *Homo erectus* populations. This would mean that modern Chinese, for example, are directly descended from Chinese *Homo erectus*, and modern Europeans are similarly evolved from European *Homo erectus*, rather than being descendants of *Homo sapiens* who migrated from Africa. In the multi-regional scheme a modern

European and a modern Chinese would have last shared a common ancestor at least one million years ago, while in the 'Out of Africa' scenario they would be linked very much more recently.

What the mitochondrial gene tree did was to introduce an objective time-depth measurement into the equation for the first time. It showed quite clearly that the common mitochondrial ancestor of *all* modern humans lived only about 150,000 years ago. This fitted in very well with the 'Out of Africa' theory and was enthusiastically welcomed by its supporters. But it came as a severe shock to the multi-regionalists. If all modern humans were related back to a common ancestor as recently as 150,000 years ago, they could not possibly have evolved in different parts of the world from local populations of *Homo erectus* that had been in place for well over a million years. Though the multi-regionalists, being thoroughly modern humans themselves, have refused to accept defeat, the mitochondrial gene tree dealt a wounding blow to their theory from which it has not yet recovered.

For us, it was great news. Mitochondrial DNA was catapulted by this controversy into its position as the prime molecular interpreter of the human past. A surge of research effort was bound to follow in laboratories all over the world. And that meant there would be lots of data with which we could compare our own results. If we were going to put the results from the old bones into a modern context, then we could not do better than use mitochondrial DNA.

4

THE SPECIAL MESSENGER

Mitochondria are tiny structures that exist within every cell. They are not in the cell nucleus, the tiny bag in the middle of the cell which contains the chromosomes, but outside it in what is called the cytoplasm. Their job is to help cells use oxygen to produce energy. The more vigorous the cell, the more energy it needs and so the more mitochondria it contains. Cells from active tissues like muscle, nerve and brain contain up to one thousand mitochondria each.

Each mitochondrion is enclosed within a membrane. Arranged in an elaborate structure within the membrane are all the enzymes required for the final stage of aerobic metabolism. This is the part where the fuel we take in as food is burnt in a sea of oxygen. There are no flames and all the oxygen is dissolved, but it is as much a piece of combustion as what happens in a gas fire or a car engine. Fuel and oxygen combine to produce energy. Fires and engines produce their energy as heat and light. Mitochondria do not give off light when they burn fuel but they do heat up – it is partly the heat given off by mitochondria that keeps us warm.

However, the main output is a high-energy molecule called ATP, which is used by the body to run virtually everything, from the contraction of heart muscles, to the nerves in your retina that is reading this page, to the cells in your brain that are interpreting it.

Buried right in the middle of each mitochondrion is a tiny piece of DNA, a mini-chromosome only sixteen and a half thousand bases in length. This is minuscule compared to the total of three thousand million bases in the chromosomes of the nucleus. Finding DNA in mitochondria at all was a big surprise. And it is very peculiar stuff. For a start, the double helix of this DNA is formed into a circle. Bacteria and other micro-organisms have circular chromosomes, but not complex multi-cellular organisms and certainly not humans. The next surprise was that the genetic code in mito-chondrial DNA is slightly different from the one that is used in the nuclear chromosomes. Mitochondrial genes hold the code for the oxygen-capturing enzymes that do the work in mitochondria. However, many genes that govern the workings of the mitochondria are firmly embedded within the chromosomes of the nucleus.

How did this all come about? The current expla-nation is stunning. It is thought that mitochondria were once free-living bacteria that, hundreds of millions of years ago, invaded more advanced cells and took up residence there. You could call them parasites, or you could call their relationship with the cells symbiotic, with both cells and mitochondria doing something for each other. Cells got a great boost from being able to use oxygen. A cell can create much more high-energy ATP from the same amount of fuel using oxygen than it can without it. For their part, the mitochondria evidently found life within the cell more comfortable than outside. Very slowly, over millions of years, some of the mitochondrial genes were transferred to the nucleus, where they remain. This means mitochondria are now trapped within cells and could not return to the outside world even if they wanted to. They have become genetically institutionalized. Even now you can see the evidence of gene transfers between mitochondria and nucleus that didn't work out. The nuclear chromosomes are littered with broken fragments of mitochondrial genes that have moved across to the nucleus over the course of evolution. They can't do anything because they are not intact. So they just sit there, as molecular fossils, a reminder of failed transfers in the past.

There is something else which is unique to mitochondria. Unlike the DNA in the chromosomes of the nucleus, which is inherited from both parents, everyone gets their mitochondria from only one parent – their mother. The cytoplasm of a human egg cell is stuffed with a quarter of a million mitochondria. In comparison, sperm have very few mitochondria, just enough to provide the energy for swimming up the uterus as they home in on the egg. After the successful sperm enters the egg to deliver its package of nuclear chromosomes it has no further use for the mitochondria, and they are jettisoned along with the tail. Only the sperm-head with its package of nuclear DNA enters the egg. The plump, fertilized egg now has nuclear DNA from both parents, but its only mito-chondria are the ones that were in the cytoplasm all along – and they all come from the mother. For that simple reason, mitochondrial DNA is always maternally inherited.

The fertilized egg divides again and again, forming first an embryo, then a foetus, which in turn becomes a new-born baby and, eventually, an adult. Throughout this process, the only mitochondria to be found are copies of the originals from the mother's egg. Though both males and females have mitochondria in all their cells, only women pass theirs on to their offspring because only women produce eggs. Fathers pass on nuclear DNA to the next generation, but their mito-chondrial DNA gets no further.

Changes to DNA, both in the mitochondria and in the nucleus, arise spontaneously as simple mistakes during the copying that accompanies cell division. Cells have error-checking mechanisms which correct most mistakes, but a few escape this surveillance and get through. If these mutations occur in cells that go on to produce eggs or sperm, known as the *germline cells*, then they can be passed on to the next generation. Mutations that occur in the other body cells, called *somatic cells* – the ones that aren't going to produce germline cells – will not be passed on. Most DNA mutations have no effect at all. Only very occasionally, when they strike and disable a particularly important gene, will mutations be noticed. In the worst cases, these mutations can produce serious genetic diseases, some of which we shall encounter in a later chapter, but most of the time they are harmless.

The rate at which mutations occur in nuclear DNA is extremely low – roughly, only one nucleotide base in one thousand million will mutate at every cell division. Mitochondria, on the other hand, are not quite so vigilant with their error-checking and allow through about twenty times as many mutations. This means that many more changes are to be found in mitochondrial DNA than in the equivalent stretch of nuclear DNA. In other words, the 'molecular clock' by which we can calculate the passage of time through DNA is ticking much faster in the mitochondria than in the nucleus. This makes mitochondria even more attractive as a tool in investigating human evolution. If the mutation rate were very low, then too many people would have exactly the same mitochondrial DNA and there wouldn't be enough variety to tell us anything much about developments over time.

There is yet another bonus. Although mutations are found all round the mitochondrial DNA circle, and this whole range was used by Allan Wilson and his students in 'Mitochondrial DNA and human evolution', there is a short stretch of DNA where mutations are especially frequent. This section, about five hundred bases in length, is called the *control region*. It has managed to accumulate so many mutations because, unlike the rest of the mitochondrial DNA, it does not carry the codes for anything in particular. If it did, then many of the mutations would affect the performance of the mitochondrial enzymes. This does sometimes happen when mutations hit other parts of the mitochondrial DNA outside the control region; there are some rare neurological diseases which are caused by mutations in genes that disable essential parts of the mitochondrial machinery. Because they are so damaged, these mitochondria do not survive well and are only very rarely passed on to the next generation. So these mutations gradually die out. The control region mutations, on the other hand, are not eliminated, precisely because the control region has no specific function. They are neutral. It appears that this stretch of DNA has to be there in order for mitochondria to divide properly, but that its own precise sequence does not matter very much.

So here we have the perfect situation for our research: a short stretch of DNA that is crammed full of neutral mutations. It would be much quicker and cheaper to read the sequence of the control region, just five hundred bases, than the entire mitochondrial DNA sequence at over sixteen thousand bases. But was the control region going to be stable enough to be useful in examining human evolution? If the control region were mutating back and forth at a great rate at every generation, then it would be extremely difficult to make out any consistent patterns over the course of longer time spans. We knew already from the work of Allan Wilson that if we were going to dig down deep into the genetic history of our species, *Homo sapiens*, using mitochondrial DNA, we needed to cover at least 150,000 years of human evolution – say 6,000 generations at twenty-five years per generation. If mutation in the control region were too frantic or erratic, it would be very hard, if not impossible, to distinguish the important signals from all the incidental, irrelevant changes after a few generations. We needed a way of testing this before embarking on the time-consuming and expensive commitment of a large study of human populations. How could we best do this?

Ideally, I wanted to find a large number of living people that could be proved to be descended through the female line from a single woman. In the course of my medical genetics research on inherited bone disease, I had worked with several large families; so now I took out the charts on which I had recorded their pedigrees. Although these went back several generations, there were depressingly few continuous maternal lines connecting the living members of these families. I could ask for the families' help to put me in touch with relatives who were not shown on the charts; but it would be a long business. Still, there seemed nothing else for it, and I began to dig out their names and addresses. On my way back home that night, while I was thinking about something else, I experienced one of those rare moments when an idea suddenly arrives from the recesses of the mind, goodness knows how, and you know within a millisecond that it is the answer to your problem, even though you haven't had time to work out why. I suddenly remembered the golden hamster.

When I was a small boy, I read in a children's encyclopaedia that all the pet golden hamsters in the world were the descendants of just one female. I can definitely say that I had not thought about this again over the intervening decades. And yet the idea surfaced now. I do remember thinking at the time that the story couldn't possibly be true. But what if it were? This would be the ideal way to test out the stability of the control region. All the golden hamsters in the world would have a direct maternal line back to this 'Mother of all Hamsters'. It follows that they would also have inherited their mitochondrial DNA from her, since it is passed down the female line in hamsters just as it is in humans. All I had to do was collect DNA from a sample of living hamsters and compare their control region sequences. I didn't need to have an accurate pedigree, because if there really had been only one female to start with they all had to trace back to her anyway. If the control region was going to be stable enough to be any use to us, then its sequence should be the same, or very similar, in all living hamsters.

I asked Chris Tomkins, an undergraduate student who, in the summer of 1990, had just started his final year genetics project in my laboratory, to see what he could find out about the golden hamster. The first thing he discovered is that, properly speaking, they are not called golden hamsters at all but Syrian hamsters. Chris went straight down to the Oxford public library and came back with some good news: he had found out that there was a National Syrian Hamster Council of Great Britain. He called the secretary and next day we were on our way to an address in Ealing, west London. Here we were greeted, with no little suspicion, by the secretary of the Syrian

Hamster Club of Great Britain – Roy Robinson (now sadly deceased).

The late Mr Robinson was the product of a vanished age, a self-taught amateur scientist of great distinction. His dimly lit study was full of books on animal genetics, many of them written by himself. He pulled out his book on the Syrian hamster. His eyesight was very poor, and even with the help of very thick spectacles he needed to hold the text right up close to his face. He confirmed the story I had read as a boy. Apparently, in 1930 a zoological expedition to the hills around Aleppo (now Halab) in north-west Syria had captured four unusual small golden-brown rodents, one female and three males, and taken them back to the Hebrew University in Jerusalem. They were kept together, and the female soon became pregnant and gave birth to a litter. There was clearly going to be no difficulty in breeding them in captivity. The university began to distribute them to medical research institutes around the world, where they became popular as an alternative to the more usual rats and mice – though they were tricky lab animals, active only at night, bad-tempered and prone to bite their handlers (good for them!). The first recipient was the Medical Research Council institute at Mill Hill in north London, which passed some on to London Zoo. By 1938 the first golden hamsters had reached the United States.

Sometimes, lab animals that are no longer required are taken home by staff and kept as pets rather than being killed. Over time, hamsters spread from one household to another and, as their popularity increased, commercial breeders added them to their catalogues and groups of hamster enthusiasts started up. In 1947 a piebald hamster appeared in one breeding colony – the first of many coat colour varieties, caused by spontaneous mutations in the coat colour genes, it showed itself because of the inbreeding within the colony. It wasn't difficult to mate the mutants with each other and produce a pure-bred strain. Breeders became ever keener to find new coat colours, and over the next few years many different such mutants were discovered and pure-bred strains established – cream, cinnamon, satin, tortoiseshell and many more. Hamsters made good pets and the availability of strains with different coats only added to the interest. Thus began the population explosion: today there are over three million hamsters kept as pets all over the world.

Mr Robinson lived in an old horticultural nursery, which at the time we visited was quite run down. A long, rectangular plot enclosed by walls of beautiful old brick contained overgrown flower beds and a handful of greenhouses with cracked and broken panes. There were also two substantial sheds, and we made our way to the first one on the left, where Mr Robinson unlocked the door to let us in. We could not believe our eyes. Inside were rack upon rack of cages, all labelled and numbered, within each of which nestled a family of hamsters. Mr Robinson had collected an example of every single coat variety that had ever been produced, and was interbreeding them to unravel the genetics. There were pure-white hamsters, lilac hamsters, hamsters with short dark fur and hamsters with long fine coats like an angora goat. So eminent was Mr Robinson in the world of Syrian hamsters that each time a new coat mutant was discovered, a pair would be sent to Ealing. We were looking at the world reference collection. To cap it all, he opened an old 'Quality Street' sweet tin and there inside, neatly stacked, were the dried skins of the original animals that had been sent to him. Martin Richards, who had made the trip along with Chris and myself, was so taken that he bought two hamsters from a pet shop in Ealing on the way home. He kept them in his flat for two years until they passed away. Of more immediate significance, we took away from Mr Robinson's collection a few hairs taken from each strain.

Mr Robinson had also given us the contact details of Syrian hamster breeders' and owners' clubs throughout the world, and Chris was about to write to them asking for hair samples when it occurred to us that this might not go down very well. We had already discovered that you needed quite a number of hairs to get out the DNA. Hamster hairs were very fine and tended to break off above the root. Although the animals didn't mind a few hairs being plucked, they were likely to feel a little uncomfortable, and so were their owners, if we asked for substantial tufts. That's when we realized we needed another source of DNA. We hit on what seemed at first a completely wild idea. We knew the DNA amplification reaction was exquisitely sensitive, which is why it had worked with the ancient DNA from the archaeological bones. Would there be enough hamster cells shed from the walls of the large intestine to survive in their droppings? Surely, not even the most devoted owner would begrudge parting with a few droppings for the cause of science. But would it work? There was only one way to find out – so next day Martin appeared with a fresh crop from his house guests. They were dried and shrivelled, rather like mouse droppings, and totally inoffensive. Even so, Chris used tweezers to pick them up and put them into a test tube. He boiled the droppings for a few minutes, spun down the sediment in a centrifuge and took a drop of the clear liquid into the DNA amplification reaction. It worked a treat.

For the rest of the summer small packets arrived from hamster enthusiasts all over the world. With their characteristic rattle, we knew immediately what they were. We eventually got DNA from thirty-five hamsters, and it wasn't long before Chris had sequenced the mitochondrial control region in all of them. They were all absolutely identical. So the story was true after all. All the pet hamsters in the world really do come from a single female. But more importantly for us, the control region had remained completely stable. From that very first hamster captured in the Syrian desert to its millions of great-great-great . . . great-grandchildren from every corner of the world, the control region DNA had been copied absolutely faithfully with not even a single mistake.

It was an amazing thought. Going flat out, hamsters can manage four or five generations a year. At that rate there would have been time for at least two hundred and fifty hamster generations since 1930. Even though all thirty-five of our hamsters would not have traced independent maternal lines all the way back to 1930, the fact that there were absolutely no DNA sequence differences between any of them had to mean that the anxiety I had that mutations in the control region might be happening too quickly was unfounded. Quite the reverse, in fact: this was a very reliable region of DNA after all, not given to fickle fits of mutation that would make it impossible to trace over the hundreds of generations we wanted to explore in our own human ancestors. Of course, there was a chance that even though the control region was stable in hamsters, it might not be in humans. I didn't think this was very likely, given the very fundamental nature of mito-chondria, and I was prepared to take that risk.

I was not alone in my interest. Before very long it was plain that other scientists were thinking along the same lines and had realized the potential of this very special piece of DNA to illuminate not only the grand schemes of human evolution but much more recent mysteries as well.

5

THE TSAR AND I

In July 1991 the remains of nine bodies were dug out of a shallow grave in birch woods just outside Ekaterinburg, formerly Sverdlovsk, in the Russian Urals. This exhumation was the culmination of years of research and persistence by the Russian geologist Aleksandr Avdonin, who thought he had located the resting place of the last of the Romanovs, the imperial Russian royal family. The last Tsar, Nicholas II, his wife, the Tsarina Alexandra, and their five children had been executed, or murdered – depending on your point of view – along with their doctor and three servants in the basement of the house in which they had been kept prisoner by the Bolsheviks. This was the night of 16 July 1918, in the turmoil of revolutionary Russia, and rather than risk the royal captives being released by White Russians who were then closing on the town, the decision was made, at the highest level, to kill them.

According to a contemporary account, the family were already in bed when the final elements of the plan were put into action. A telegram was sent to Lenin in

Moscow asking him to sanction the execution. Delays on the way meant that it did not reach the Kremlin until after eleven o'clock at night. The reply which gave the green light arrived at one o'clock the following morning. At half past one, a truck drew up at the house ready to take away the bodies. The family were roused and informed that, because of the military action in the town, they must spend the rest of the night in the basement where they would be safer. The Romanovs had heard the distant sound of artillery every night for the past fortnight, and saw nothing particularly sinister in this request, so they all made their way quietly down the stairs.

When they got down to the cellar, they were still not alarmed to find several guards had joined them. Even when they were asked to line up in a group, they were not suspicious. Then the leader of the execution squad approached the Tsar and took a piece of paper out of his pocket with one hand while his other rested on a revolver inside his jacket. Hastily he read the notice which condemned them to death. The Tsar was confused. He turned to his family, then to the guards, who drew their weapons. The girls started to scream. The firing began. First to be hit was the Tsar; he slumped to the floor. The cellar echoed with the screams of the victims mixed with the sound of gunfire and bullets as they ricocheted around the room. It was pandemonium, and the room soon filled with smoke, making it even harder for the squad to pick out their targets who were rushing to and fro in a blind panic. The order to cease firing was given and the victims were finished off with bayonets and rifle butts. It had taken less than three minutes to put an end to a dynasty that had ruled Russia for three hundred years.

The house is no longer there. It was demolished in 1977 on the orders of the First Secretary of the Sverdlovsk Region, the young Boris Yeltsin. But the fate of the Romanovs themselves remained something of a mystery. In the atmosphere of uncertainty and disinformation that prevailed in Bolshevik Russia, just because there were official reports of events, even events as historic and infamous as the 'execution' of the Romanovs, this did not mean that the events described had actually taken place. There were persistent rumours, actively encouraged by Soviet propaganda at the time, that the Romanovs had been sent to a safe place for their own protection. Another rumour circulated that the Tsarina and the children had been smuggled out to Germany. Yet another had the Tsar in the Kremlin, where Lenin was preparing to reinstate the monarchy as soon as the bourgeoisie had been eliminated and the Tsar was 'reconnected to the people'.

The discovery of the skeletons at least promised to introduce some objective evidence into the debate. The proof of the execution story depended entirely on showing that the remains taken from the pit really were those of the Romanovs. The location at least tallied with some contemporary accounts that told of the bodies being loaded on to a truck and driven into the woods on the outskirts of the town. According to these accounts, the executioners panicked when their truck became stuck in the mud, and they threw the bodies into a hastily dug pit before dousing them with sulphuric acid in a vain attempt to remove all features which could be used for identification.

When all the recovered bones were assembled, it soon became clear that these were the remains of only nine bodies, two fewer than there should have been if all the victims of the massacre had been buried in the same grave. After the long and painstaking process of refitting more than eight hundred bones and rebuilding the shattered skulls that had been crushed by the rifle butts of the burial detachment, it was concluded from the skeletons that the nine bodies were those of the Tsar and Tsarina; three of their five children – Maria, Tatiana and Olga; their physician, Dr Eugeny Botkin; and three servants, Alexei Trupp the valet, Ivan Kharitonov the cook and Anna Demidova the Tsarina's maid. There was no sign of the bodies of the youngest daughter Anastasia, nor of the Crown Prince, the Tsarevich Alexei. Other than these reconstructions, what further tests could be done on the remains to confirm their identity?

We had already published a paper in 1989 showing that DNA could be extracted from much older bones than these, so it was only natural to try to get DNA from the Ekaterinburg remains in the hope of confirming that these were the Romanovs. The work was carried out by the Russian Academy of Sciences and the British Forensic Science Service. First they used conventional forensic genetic fingerprints to identify the sex of the skeletons and to confirm that they did indeed include a family group of two parents and three children. DNA from the remains presumed to be those of Dr Botkin and the servants showed that they were unrelated to the family group or to each other. So far, everything fitted in well with the conclusions of the bone experts.

These scientists also succeeded in recovering mitochondrial DNA from the bones, and came up with two different sets of sequences from the family group. The female adult, the presumed Tsarina, and all three children had an identical mitochondrial DNA sequence. The male adult in the family group, the presumed Tsar, had a different sequence. This was exactly what you would expect from a family. All three children had inherited their mother's mitochondrial DNA sequence while the father, who had got his from his own mother, had not passed it on to any of his children. However, on its own, extracting the mito-chondrial DNA and sequencing it did not identify this family as the Romanovs – any family would show the same pattern of identity between mother and offspring, with the father showing a different sequence. The only way of proving *which* family this was was to locate living relatives of the Tsar and Tsarina who were con-nected to the dead Russians through a series of entirely maternal links. They didn't have to be especially close relatives; the real power of mitochondrial DNA is that it is not diluted by distance. So long as the connec-tions are exclusively maternal and not disrupted by a father–child link, then the mitochondrial DNA will be identical.

Fortunately, it was possible to trace living direct maternal relatives of both the Tsar and the Tsarina. The Tsar had an unbroken maternal connection through his grandmother Louise of Hesse-Cassel, the Queen of Denmark, to a Count Nicolai Trubetskoy, seventy years old and living in peaceful retirement on the Côte d'Azur after a lifetime as a merchant banker. The Tsarina could trace a direct maternal link through her sister Princess Victoria of Hesse to His Royal Highness Prince Philip, the Duke of Edinburgh, the husband of Queen Elizabeth II. After several rounds of discreet negotiation both men agreed to provide a small blood sample from which their DNA could be extracted. What would they show?

The notation everyone uses to compare mitochondrial DNA sequences involves quoting differences from a set reference sequence, in fact the very first mitochondrial DNA to be entirely sequenced, by a team from Cambridge in 1981. In this notation, a DNA sequence which differs from the reference sequence at the fifteenth and one hundredth positions in the 500 base control region segment is abbreviated to 15, 100. The sequence from the Duke of Edinburgh was 111, 357 using this notation. At all the other 498 positions along the 500 base stretch, the Duke's sequence was exactly the same as the reference sequence.

It is always much harder to get a complete sequence in one go from ancient DNA than from a modern sample. The strands are fragmented by the ageing process, so even the relatively short 500 base segment of the control region has to be built up in overlapping stages of a hundred bases or so. This is a laborious process, but eventually the sequences of the presumed Tsarina and her three children were typed. They all had exactly the same sequence of 111, 357. They were all an exact match with the Duke of Edinburgh.

The same, however, was not true for the adult male, the presumed Tsar. He was not an exact match with Count Trubetskoy. Whereas the Trubetskoy sequence was 126, 169, 294, 296, the presumed Tsar's DNA had mutations at only 126, 294 and 296 – very similar but not identical. This was a definite setback. There was so much circumstantial evidence connecting the bodies with the Romanovs, and there was the exact match of the females with the Duke of Edinburgh. But there is no point in doing a genetic test if you don't take notice of the result. A close match is not an exact match. And if the maternal connection over six generations with Count Trubetskoy was unbroken, the match would be exact.

Was there a chance that the Count was not really a relative of the Tsar, even though the family tree had recorded him as such? If so, there would have to have been a break somewhere along the line going back from the Tsar to Louise of Hesse-Cassel and then down to Count Trubetskoy. It would mean, in fact, that one of the people on this line had a different mother from the woman recorded on the pedigree. This is always a possibility – there could have been an adoption or a mix-up at the birth – but these are only remote possibilities. If it were a paternal line that was being followed it would be different. A child can easily have a different biological father from the man married to his mother; but such mis-identification is much more unlikely down the maternal line. After all, both mother and baby have to be present at the birth. The only formal conclusion that could be reached was that this was not the Tsar; and so, that since the conventional genetic fingerprints had already identified him as the father of the three children found in the grave, this was not the grave of the Romanovs after all.

But even though the mitochondrial DNA sequences of Trubetskoy and the male skeleton were not exactly the same, they were very close; and so near a miss invited further thought. They both shared three mutations at positions 126, 294 and 296. Trubetskoy had another one at position 169. Was it possible that there had been an error in reading the sequence of the 'Tsar's' mitochondrial DNA? The team went back to the original trace from the sequencing machine and looked very closely at the readings at position 169 for the 'Tsar's' sample. The trace itself looks like four superimposed lines of different colours, representing the readout from four separate channels which detect the four DNA bases: red for T, black for G, blue for C and green for A. While Trubetskoy's trace showed a clear red peak at position 169 corresponding to the mutation T, the 'Tsar's' trace at the same position showed the blue peak for C, the same as the reference sequence. But underneath the blue peak was a small red blip. Could it be that the 'Tsar's' DNA was a mixture of two mitochondrial DNA sequences, the main one with the sequence 126, 294, 296 and another, much smaller, with the same sequence plus the mutation at position 169? There was one way to find out, and that was to clone it.

Cloning is the only way to separate the different DNA molecules in a mixture. Briefly, it involves tricking bacteria into accepting just a single molecule of DNA and then copying it as if it were their own. Getting DNA into bacteria is a very inefficient process; only one in a million accepts it. Still, if just a couple of dozen bacteria can be persuaded to take in the DNA, they can be treated in such a way that the only bacteria to survive and grow as colonies on a culture dish are the ones with the extra DNA. They can then be picked off and the DNA sequenced. Within each colony, all the DNA will be copies of the original molecule that was accepted. If there is a mixture of two different DNA molecules to start with, some of the colonies will have one type and some will have the other. The scientists managed to create twenty-eight clones containing mitochondrial DNA from the 'Tsar'. When each of these was individually sequenced, twenty-one contained the main sequence 126, 294, 296 read from the original trace, without the mutation at 169. But the DNA from seven clones did contain the additional 169 mutation, making it absolutely identical to Count Trubetskoy's.

What the researchers had stumbled across was the very rare state where a new mutation, in this case at position 169, is part way to becoming established. This state, formally called *heteroplasmy*, had scarcely ever been observed before and was very little understood. As we will see in a later chapter, we know a lot more about heteroplasmy now; in 1994, when the paper on the 'Romanov' remains was published, it was a novelty. But it did get the researchers off the hook. Here was the evidence they needed that there was indeed a continuous maternal link between the bones of the Ekaterinburg 'Tsar' and a living relative of Tsar Nicholas II.

The mitochondrial DNA matches were certainly good evidence to support the case that the Ekaterinburg bones were the remains of the Romanovs. But was it proof? Proof can never be absolute. It is always relative. In the case of the Romanovs the degree of certainty could be given a mathematical form depending on how common these mitochondrial sequences are in Europe. In those early days of the research we didn't know many European sequences, so it was hard to know how strong the evidence was. Now we have far more sequences to compare, and we know that the Duke of Edinburgh's sequence (111, 357) is actually extremely rare: it has not been found again in over six thousand Europeans. Since it has not been seen elsewhere, we cannot accurately estimate its frequency, but it is very unlikely to be higher than one in a thousand. This means there is, at most, a one in a thousand chance that the mitochondrial DNA sequence from a European picked at random would match the Duke of Edinburgh. So there was still a very small chance that the female Ekaterinburg bones did not belong to the Tsarina and her children at all, but to another family who just happened to have the same mitochondrial DNA as the Duke of Edinburgh. The Trubetskoy sequence (126, 169, 294, 296) is again very rare and has not been seen in six thousand modern Europeans. However, the Tsar's main sequence (126, 294, 296) is much more frequent, with just under one in a hundred Europeans matching it exactly. So, once again there was a small but finite chance that the bones of the adult male were not the Tsar's but those of someone else who just happened to match.

Though the DNA matches gave a pretty high level of proof already, there is a further level to consider. We have not yet taken into consideration the fact that the two sets of matching sequences were found in the same grave and came from the parents of the three children, according to the DNA fingerprints. How does that affect the result? The answer is that it makes the level of proof that these really were the Romanov bones very high indeed. The probability of getting matches to *both* sets of mitochondrial DNA sequences just by chance is the mathematical product of the individual probabilities. That is one in a thousand multiplied by one in a hundred, which comes to the vanishingly small figure of one in a hundred thousand. Add to that the circumstantial evidence which led to the discovery of the grave and the evidence of bullet wounds, and the proof climbs even higher towards 100 per cent.

But one mystery remained. Only five Romanov bodies were ever found – two adults and three girls. Formally, one might take the view that this is evidence against the remains being those of the Romanovs at all. But it tallies with persistent rumours that some of the children had escaped execution. A Soviet

96

announcement that only the Tsar himself had died and that the rest of the family had been sent to a safe place was swiftly followed by the appearance of all too obvious impostors. For a while, every town in Siberia, then in the hands of the White Russians and not the Bolsheviks, had its own 'Grand Duchesses' and 'Crown Princes'. Most were obvious frauds, but some managed to do well out of the deception for a while. One enterprising businessman even ran a regular export service, persuading local millionaires to part with their cash to help him send the imperial refugees abroad to safety. His accomplice, playing the part of one or other rescued 'Grand Duchess', even allowed the entranced sponsors to kiss her hand as she bade a tearful last farewell to her beloved country.

The Tsar's mother, the Dowager Empress Marie Fedorovna, exiled in Copenhagen, did more than anyone to keep alive the myth that her family had survived, refusing to accept that they had died right up to her own death in 1928. Throughout the last ten years of her life she was challenged to accept the claims of numerous pretenders as her grandchildren. By far the most persistent of these claims was that of the woman who became known as Anna Anderson. It began when a young woman jumped from a bridge into the Landwehr canal in Berlin in February 1919, seven months after the Ekaterinburg massacre. She was rescued, but resolutely refused to reveal her identity and was confined to a mental hospital as 'Fräulein Unbekannt' – the unknown lady. One of her fellow inmates, Clara Peuthert, became convinced, from an account of the massacre in a Berlin newspaper, that this withdrawn and uncommunicative patient was none other than the Grand Duchess Tatiana, the second of the Tsar's four daughters. After she was released from the asylum, Clara Peuthert championed Fräulein Unbekannt's case among the White Russian émigrés in Berlin. Using these contacts, she arranged a visit by the Tsarina's former lady-in-waiting, Baroness Buxhoeveden. This was the first of many often disastrous encounters with people anxious to establish the real identity of the 'survivor' that went on for most of the rest of her life. On this occasion, Fräulein Unbekannt hid under the bedclothes. The redoubtable Baroness pulled the sheets aside and dragged her out of bed. She could not possibly be Tatiana, exclaimed the Baroness. She was far too short. This rather obvious disqualification only made Fräulein Unbekannt declare that she had not actually said she was Tatiana, who was, in fact, the tallest of the Tsar's daughters. At only 5ft 2in, Fräulein Unbekannt was much more the size of Anastasia. And so that is who she claimed to be for the rest of her life, taking the name Anna as an abbreviation of Anastasia and adding Anderson many years later to confuse local journalists during her stay at a hotel on Long Island, New York.

Anna Anderson's pathetic life, spent in hospitals and the homes of her supporters, came to an end in 1984 near Charlottesville in Virginia. If she were Anastasia she would have been eighty-three years old. Over the years she became embroiled in unending legal battles between her supporters and those who wanted her claim dismissed. Her opponents were accused of wanting to prove the death of the Tsar's entire family so that they could benefit from money the Romanovs had deposited in overseas bank accounts; her supporters were accused of coveting those fortunes for themselves. Throughout all this conflict and controversy, Anna Anderson herself never vigorously prosecuted her claim. Whenever there was a chance to impress one of the Tsar's relatives who had been persuaded to visit her, she would be untalkative and uncooperative, refusing to answer questions and often hiding in her room. While this behaviour annulled her claim in the eyes of her detractors, it was her very reluctance to press her case, coupled with an absolute self-belief that she was the Grand Duchess Anastasia, that convinced her supporters. The matter was never conclusively settled during her lifetime, and she passed away with her claim neither validated nor disproved. Fortunately for her, she died before the cold eye of genetics could be turned on the case. If she had lived another few years, like her contemporary Queen Elizabeth, the Queen Mother, who is still alive at the age of 101, then her lifetime of deception would have been mercilessly exposed.

In a thrilling piece of detective work, mitochondrial DNA was recovered from a stored biopsy from Anna Anderson, taken when she was in hospital for an operation to remove a bowel obstruction in 1979. It had a completely different sequence from the Tsarina's DNA. Anna Anderson could not possibly have been Anastasia. A test that had taken a month at most to perform had demolished at a stroke one of the most enduring and romantic sagas that had gripped the world from one end of the twentieth century to the other. Such is the power of DNA to dispel myths – even those we might prefer to have believed.

The sequence from Anna Anderson's biopsy did, however, match a living maternal relative of one Franziska Schanzkowska, a patient in a Berlin mental home who disappeared in 1919 shortly before 'Anastasia' made her appearance in the same city. Opponents of Anna Anderson's claim had always suspected her to be Franziska Schanzkowska, and not Anastasia as she claimed. DNA proved them right.

So the mystery of Anastasia lives on. In our laboratory we have more than once been asked to examine the DNA of further claimants. Sadly, none of them has passed the scrutiny of the DNA test. In the 1956 film *Anastasia*, written as a romantic fiction rather than as a true record of events, the Dowager Empress Marie Fedorovna puts Anastasia, played by Ingrid Bergman, through a series of tests to prove whether she is her grand-daughter. She eventually accepts the young woman, and the film ends happily. It would not have, had DNA been around. But the film also brought its own reward for Anna Anderson, who received a share of the royalties.

If Anna Anderson, the most convincing of the claimants, was not Anastasia, perhaps the Grand Duchess had perished with her sisters after all. The pit contained the bodies of only three girls. Two bodies, those of one of the Grand Duchesses and the Crown Prince the Tsarevich Alexei, are still missing. Alexei, too, has had his impersonators. A Soviet sailor, Nikolai Dalsky, persisted in his claim to the imperial crown, which in Soviet Russia showed a certain self-confidence, until he died in 1965. His son, 'Nikolai Romanov', inherited the claim on his father's death and refers to his own son Vladimir as the Tsarevich. However, the truth is almost certainly that the whole family were killed. Written reports, for what they are worth, record that the men whose job it was to dispose of the bodies first tried to burn them in the woods near the site of the pit where the remains were found. They built a pyre and put on it first the smallest body, that of Alexei, then one of the Grand Duchesses, doused them in petrol and set fire to them. But the flames did not consume everything. Teeth and fragments of bones lay near the fire. The plan was changed and the rest of the bodies were flung into the shallow pit. If that version of events is true, the last remains of Alexei and Anastasia lie not in the graves of the pretenders but charred and burnt beneath the leaf litter of a wood in the Russian Urals.

Though I like the odd vodka, I have never considered myself a Romanov; but I couldn't help noticing that my own DNA sequence matched that of Tsar Nicholas II. If we ignore for the moment the minor component of the Tsar's DNA introduced by hetero-plasmy at position 169, we both have the notation 126, 294, 296. Does it mean that I am related to the Romanovs, even distantly? The amazing answer is 'yes'.

This is the point to stop and take in one entirely logical yet utterly extraordinary fact which forms the basis of a lot of what this book is about. If any two people trace their maternal line back – through mother, grandmother, great-grandmother and so on – eventually the two lines will converge on one woman. If the two people are brothers or sisters, then it is simple: their maternal lines meet in their mother. If the two people are cousins, the children of two sisters, then the lines converge on their shared maternal grandmother. Even though most people who have not researched their family trees will lose the trail not far beyond that, the principle is maintained no matter how long you go back into the past. Any two people, in your family, your town, your country – even the whole world – are linked through their mothers and their mothers' mothers to a common maternal ancestor. The only difference between any two people is this: How long ago did this woman live?

Further than a handful of generations back, the written records of most maternal connections are completely lost to us, so we just would not know the answer to this question. But the DNA doesn't forget. The mitochondrial DNA, because of its special inheritance exclusively through the female line, traces exactly that path back in time. And because the sequence of mitochondrial DNA changes due to random mutations, albeit very slowly, we can use it as a sort of clock. If two people shared a common maternal ancestor in the recent past, then their mitochondrial DNA will not have had time to change through mutation. Like the hamsters, their mitochondrial DNA sequences will be exactly the same. If she, the common ancestor, lived further back in time then there is a chance that a mutation will have occurred somewhere along one or both of the two tracks which lead back to her from the present day. If she lived further back still, there might be two or more mutations. By counting the differences between the two sequences, we can estimate the length of the matrilineal connection between any two people in the world. To put dates on to this we need to know the mutation rate for mitochondrial DNA. We will look in greater depth at how the rate is estimated in a later chapter (see pp. 196–9). The best estimates are that, on average, if two people had a common ancestor ten thousand years ago then there would be one difference in their control region sequences. If the common maternal ancestor of two people lived twenty thousand years ago, then we would expect to see two mutational differences in their mitochondrial DNA.

Of course, there is not the faintest chance of knowing from any written source whether any two people were related through their maternal lines twenty thousand years ago, so we work it the other way around. If two people have exactly the same control region sequence, their common ancestor will have lived, on average, some time in the last ten thousand years. The Tsar and I do have the same control region sequence. So our maternal ancestry, working back through, on my side, my mother Irene Clifford and her mother Elizabeth Smith and on the Tsar's, through his mother, the Dowager Empress Marie Fedorovna and her mother Louise of Hesse-Cassel, Queen of Denmark, most likely converges on a common ancestor who lived within the last ten thousand years. Not close enough for me to make a realistic claim to the Romanov fortunes, I think.

Measuring ancestral connections in tens of thousands of years may seem too crude to be interesting. However, although the mitochondrial mutation rate seems incredibly slow, it is fortunately just about right for studying human evolution over the last hundred thousand years – which is when most of the action happened. If the mutation rate had been much faster than it is, relationships would be harder to follow. If it were much slower, there would be too few differences between people to see any patterns at all. Taking the next logical step, if any two people can trace a common maternal ancestor, it follows that any *group* of people can do the same. I slowly realized that we held in our hands the power to reconstruct the maternal genealogy of the whole world. Not exactly world domination; but I'm sure my distant cousin, Nikolai Aleksandrovich, Imperial Tsar of all the Russias, would have approved. The question was: where should we begin?

6

THE PUZZLE OF THE PACIFIC

At nine-fifteen every evening Air New Zealand flight
NZI takes off from Los Angeles International Airport.
Within thirty seconds it has crossed the short stretch of
dry land between the end of the runway and the ocean.
There is no throttling back of the engines to cut down
on the noise levels. There is no need. Flight NZI is now
over the Pacific, and will not see land again until it
crosses the Coromandel peninsula on the North Island
of New Zealand as it makes its approach to Auckland.
But that is still seven thousand miles and fourteen
hours ahead. Between then and now there is only the
open ocean beneath us – the apparently endless reach of
the Pacific Ocean. Sprinkled across this vastness are
thousands of islands, but so dwarfed are they by the sea
that you are very unlikely to catch even a glimpse of
any of them from the plane. And yet, by the time the
first European ships began to explore the Pacific, every
one of these islands had been found and settled by the
people I have come to regard as the greatest maritime
explorers the world has ever seen – the Polynesians.

I would like to be able to say that my decision to

work in Polynesia was the result of careful planning, of balancing the scientific advantages of studying island populations with the difficulty and expense of working on the other side of the world. I would like to be able to say that, but the truth is that it all came about by accident – literally. In the autumn of 1990 I was taking a term's sabbatical leave and had arranged to spend part of it at the University of Washington in Seattle and the rest in Melbourne, Australia. This meant crossing the Pacific and, since I had never seen a tropical island before, I scheduled stop-overs in Hawaii and in a place called Rarotonga in the Cook Islands. I had never heard of Rarotonga, and only very vaguely of the Cook Islands for that matter, but it fitted into the flight schedules more conveniently than the better-known alternatives of Tahiti or Fiji.

It also had more by way of contrast to offer. Hawaii is certainly tropical and very beautiful, but at least around the capital, Honolulu, on Oahu there is no doubt at all that you are still very much in America with high-rise, pizza and pet cemeteries. Landing in Rarotonga is a very different cultural experience altogether. There are no luggage carousels: you just pick up your bags from a pile. A man with a guitar is singing a welcoming song as if he means it, which is impressive at four o'clock in the morning. And then there was Malcolm. Cheery and ruddy-faced, Malcolm Laxton-Blinkhorn is English, but nowhere near as grand as his name suggests. He has had what might be called a varied career – marine commando, sheep farmer, actor, television producer . . . and now hotelier in Rarotonga, having married a local girl. Although his hotel was on the beach at the other side of the island, Rarotonga being only 26 miles round it didn't take us long to get there. It was still dark, but who could resist going down to the water's edge and just sitting? Slowly I become aware that it is not as quiet as it should be. There is a distant but persistent low roar, like a busy motorway a mile or two off. But there are virtually no cars on the island and certainly no motorways. The sound I hear is the ocean. As the light grows I can make out a thin white line near the horizon. This is where the swell of the ocean, even on calm days like today, pounds into the coral reef that surrounds and protects the island.

My plan was to spend just a few days on Rarotonga before going on to Melbourne and carrying on with my work. Like most visitors I hired a small motorcycle, took my driving test, which consisted of riding 50 yards up the road and back to the police station, got my driving licence and set off. Straight into a palm tree. I broke my shoulder. I couldn't leave the island until it had set. Several weeks, I was told. So I settled in for a long stay.

Rarotonga is the main island of the Southern Cooks, a widely scattered archipelago seven hundred miles to the west of Tahiti. The islands get their name from Captain James Cook, the eighteenth-century English navigator, whose portrait (and it always seems to be the same one) is everywhere on the island, even fixing you with his inscrutable gaze as you down a bottle of Cook Islands lager. Though Cook explored many of

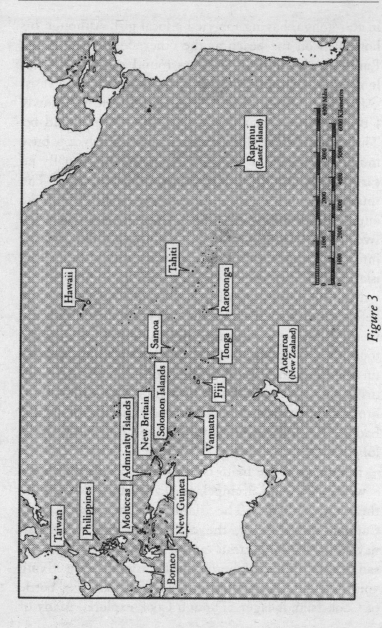

Figure 3

the islands in the group he inexplicably failed to sight Rarotonga, though it is the largest of the Cooks and rises to 650 metres. The honour of being the first Europeans to land on Rarotonga went to the mutineers of HMS *Bounty*, who in 1789 stopped on their way to the even more remote Pitcairn Island in their search for a refuge far away from the long arm of the British navy. Today the Cook Islands are internally self-governing, allied with New Zealand in foreign affairs and defence; but they were once a British protectorate and are still a member of the Commonwealth. Even though I doubt whether one in a hundred English people have ever heard of the Cook Islands, the islanders still retain some of the customs of their former colonial patrons. With a lot of time on my hands, and my arm in a sling, I went along to hear a debate in the Cook Islands parliament. The parliament building may only have been a set of corrugated iron roofed huts near the airport runway, but the procedures were every bit as formal as in the House of Commons at Westminster. At the front of the chamber sat the Speaker, through whom all remarks were addressed. Bills were introduced for first readings. Committee stages took place on the floor of the House, and full-scale debates were followed by a division. And guillotine motions. It was getting round to five o'clock in the afternoon when, with a long-winded debate on the pay of MPs and civil servants in full swing, the government introduced a guillotine motion to impose a time limit. And the reason? The cabinet had agreed to sing at the secondary school's netball team fundraiser at six-thirty, so parliamentary business had to finish by six. This was a place which had obviously got its priorities right.

Another legacy of the past was the museum and library. Even though it was surrounded by coconut palms and mango trees dripping with fruit, once inside I could have been in home counties England: silence, shelves of books and an inconspicuous librarian with a rubber stamp to frank the withdrawals. And empty. There was a substantial collection of books about the Pacific, and I began to read about the part of the world where I was an unplanned (but not too unwilling) prisoner until my fracture healed. Sitting on the fringe of the beach, staring out to the ocean beyond the crashing surf on the reef, knowing that it stretched for thousands of miles in every direction, I found one question persistently nudging me. How did the Polynesians discover and settle this island, and where had they come from?

Captain Cook, though not the first by any means, was by far the most widely travelled of the European navigators who explored the Pacific. Raised in humble circumstances in Yorkshire, and desperate to go to sea as soon as possible, he joined a ship from the port of Whitby. This was a time when an aristocratic pedigree was almost essential for a successful career in the Royal Navy; however, by his sheer brilliance in navigation Cook rose through the ranks to command his own ship. So impressive was he in his navigation of the notorious St Lawrence River during the war against the French in Quebec that he was chosen to take command of HMS *Endeavour* and lead a scientific party from the

Royal Society to observe the transit of Venus across the face of the sun. Timing this rare event was important in the calculation of the distance between the earth and the sun, and the best opportunity for observing the 1769 transit was to be found in Tahiti. This mission accomplished, Cook set out on further explorations of the Pacific which took him, in this and his other two voyages, to New Zealand, Australia, the Pacific northwest coast of America, through the Bering Straits and finally to Hawaii, where he was killed by natives at Kealakekua Bay on the Big Island on Valentine's Day 1779.

As a navigator, Cook took a professional interest in the question of the origins of the people he encountered on these remote and scattered islands. Over the period of his three voyages he observed the similarities, in both looks and language, between islanders as far apart as Hawaii, Tahiti and New Zealand, and deduced that this meant they all shared a common origin. But where was it? Polynesian tradition too speaks of an ancestral homeland, Havaiiki, but without being specific as to its location. Cook knew only too well that the winds, and the currents, of the Pacific move from east to west across the ocean, from the Americas to Asia. If Polynesians came originally from Asia, then they would have had to battle against both wind and current; if they came from the Americas, they would have been assisted on their voyage by these same natural elements – and these were considerable forces. The Spanish navigators were the first Europeans to explore the Pacific, and they could cross only one way, from east to west. Having sailed from their bases in Central America to the Philippines, they could not sail back the way they had come and had no option but to return by the Great Circle route, north past Japan and Alaska then south down the Pacific seaboard of North America. If Spanish galleons with their formidable sail power and sophisticated navigation could not defeat the winds and currents, then how could the far smaller vessels of the native Polynesians possibly have done so?

One particularly condescending group of western anthropologists were so convinced that the Polynesians were simply too incompetent to organize anything re- sembling a deliberate voyage of exploration, especially if it meant sailing into the wind, that they felt no further proof of the islanders' American origin was needed. In their view, the only possible way for these primitives to have reached the islands was by getting lost while out fishing and just drifting on to them – never mind that this would require them to have gone fishing with their whole families, their livestock and a few taro plants stowed on board. This appalling legacy of white colonial attitudes is still keenly felt by many Polynesians. Proof of their Asian origin would certainly crush this nonsense once and for all, and establish their ancestors as supreme masters of the sea.

The controversy in the minds of Europeans surrounding Polynesian origins has lasted for two hundred years. On the one hand, the evidence of archaeology and language, and the types of domesticated animals and plants found in Polynesia, all point to an origin in south-east Asia. On the other, there has been a persistent tradition, most recently revived by the Norwegian anthropologist Thor Heyerdahl, that puts the origin of the first Polynesians in the Americas. Of the evidence for an American connection, the most compelling is the widespread cultivation throughout Polynesia of the *kumara* or sweet potato, which no-one doubts is native to the Andes of South America. In his books Heyerdahl also provides other connections of language, mythology and some archaeology, like the stone facings on carvings found in Easter Island which bear a striking resemblance to the style of the Incas. But his most celebrated piece of evidence is the voyage of *Kon-Tiki*, the balsa raft that he used to sail, or rather to drift, from the coast of South America four thousand miles to the Tuamotu islands not far from Tahiti. Of course, to demonstrate that it can be done does not mean that it *was* done; but *Kon-Tiki* remains a persuasive argument to a lot of people.

Irritated by what they saw as a stunt by Heyerdahl, the serious anthropologists who had painstakingly pieced together the evidence for an Asian origin did not hide their feelings in their writings. Sitting in the library in Rarotonga, I was shocked by the toxins that dripped from the pages whenever Heyerdahl's theories were mentioned. His ideas may not have enjoyed wide support among academic anthropologists, but to me, coming fresh and ignorant to the field, his evidence taken at face value seemed to have at least some merit. How strange, I thought, that otherwise moderate and scholarly academics should suddenly lose it when the H-word was mentioned.

I sat in Lucy's café in Avarua, the capital (indeed, the only town) of Rarotonga, having an ice-cream, just looking at the people coming and going. Did they look more Asian or more American? It wasn't obvious to me. I distinctly remember one small girl who could have come straight from a *National Geographic* cover story on the Amazonian rainforests. If only I could test the mitochondrial DNA of the people in the café! I was sure I would be able to tell whether their closest genetic links were with Asia or America. So, at the next hospital appointment to review my fractured shoulder, I explained that I was a geneticist and what I had in mind. Somehow or other I managed to persuade the hospital to let me have the remnants of thirty-five blood samples left over from blood-sugar tests. Diabetes is very common in Rarotonga, and so there are a lot of tests for blood glucose levels. I stored the samples in the freezer back at the hotel. After my shoulder healed – a little too quickly, I thought – I carried these precious phials of blood with me to Australia, where I very nearly had them confiscated by customs, then eventually back to England and my laboratory.

The day after I got back, I unwrapped the samples. There was blood oozing everywhere. The glass tubes had smashed – but fortunately, not all of them. Twenty were still intact, and I got on with sequencing their mitochondrial DNA. Nowadays, DNA sequencing is done automatically in extremely expensive machines, but in the early 1990s it was a manual operation which involved tagging fragments of DNA with mildly radio-active isotopes and separating them in an electric field.

There was a point at the end of the long process when the X-ray film which recorded the pattern of radioactive bands that revealed the sequence slowly issued from the developing machine. It was always a tense moment. Will there be a good set of bands? Will there be any bands at all? If the bands are too weak or absent altogether, then something has gone wrong and it's back to the laboratory bench for another three days.

This time, with the first ten of the twenty samples, everything had worked. Drawn across the X-ray film were four wide columns of dark bands, like bar codes, where the weak radioactivity had blackened the photographic emulsion. The four columns were each divided into ten tracks, one for each sample. Each of the four columns read the sequence of one base, so by putting them together the complete sequence could be worked out. I had arranged it this way, with the ten samples side by side, so that it was easy to see where the differences between individuals were. That was what I really wanted to focus on – the differences between people, rather than the similarities. A straight line across all ten tracks meant that all ten samples were identical at that position; a line with interruptions meant that some samples were different at that position.

In the lab we had sequenced ourselves and a few friends, mostly European, and typically there would be a couple of dozen lines in each batch of ten samples that showed these tell-tale interruptions. When the Rarotongan film came sliding out of the developer there were bands all right, but there was not a single interruption. They were all exactly the same. Had I made a mistake? Had I inadvertently combined the samples somewhere along the line? I needed to develop the second film from samples 11–20 to find out. When this came out it looked at first as if I really had made a mistake. Another whole film of straight lines. But then I saw one track, one individual, that was different from all the rest. Very different. And three other tracks had a single interrupted line. So they hadn't been mixed. They were real results. I realized at once that they were stunning, and that before very long I would have the answer to the origin of the Polynesians.

Reading through the sequences more carefully and comparing them with the European reference sequence, I saw that the major sequence shared by sixteen of the twenty Polynesians was different at four positions: 189, 217, 247 and 261. The similar sequence shared by three individuals on the second film differed from this only in that they did not have the variant at 247. Otherwise their mitochondrial DNA was identical; they had to be very closely related to the first sixteen. But the twentieth sample was completely different. It had nine variants from the reference sequence along the control region, none of which was shared with the main Rarotongan cluster. Since the blood samples had come from the outpatient clinic in Avarua, there was no guarantee that they were from native Rarotongans, and so I assumed that this unusual sequence was from a tourist or a visitor from another part of the world. Since very few mitochondrial sequences had been published in 1991, there was no telling where on the globe this might be.

I concentrated on the main result – the astonishing similarity of nineteen of the twenty samples. This had to be the mitochondrial DNA of the original Polynesians. All we had to do to solve the Polynesian question was to look in both south-east Asia and South America for comparisons. If we found DNA matches in Chile or Peru, or even in coastal North America, then Heyerdahl was right. If we found them in south-east Asia, he was wrong. If we didn't find a match in either region, then everyone was wrong. Whichever turned out to be true, one thing was certain: we were going to settle, once and for all, the debate that had raged for over two hundred years. I started to plan my next trip.

You might be asking yourself at this point: 'Surely if it were as easy as that, blood groups would have given the answer long ago?' It isn't as if the blood groups of Polynesia had never been studied; indeed, the first results from Samoa in central Polynesia had been published in 1924, only five years after the *Lancet* paper by the Herschfelds which first introduced the potential of blood grouping in anthropology. The south Pacific, as I was fast learning, had been a popular place for scientific fieldwork for a long time. However, while they formed a plank in the argument in favour of a south-east Asian origin, decades of work on blood groups and other classical genetic systems had still not produced a definite answer to the puzzle, first because the variations are not definitive, and second because the evolutionary relationships between the types are not known. For example, Polynesians, native South Americans and south-east Asians all have a high frequency of blood group O. Polynesians also have quite a high frequency of blood group A, which is virtually absent in South America. But they also have a low frequency of blood group B, which is quite common in south-east Asia. So what can you make of all that? Which theory do these data support? Advocates of the Asian origin would argue that the extreme rarity of blood group A in native South Americans means that the Polynesian blood group A couldn't have come from South America. Supporters of the South American case could legitimately respond by saying, as Arthur Mourant suggested in 1976, that the Polynesian blood group A came originally not from Asia but from Europeans through intermarriage over the last three hundred years. And anyway, where's all the blood group B that should have come from Asia? Add to all this uncertainty the fact that, ultimately, all native Americans trace their origins to Asia through the settlers who crossed the Bering land bridge thousands of years before, and you have a complete mess. Blood group O could have reached Polynesia either directly from Asia or via the Americas. There is no way of knowing. With only three blood group genes – A, B, and O – certainty remains out of reach.

Other classical genetic markers are more variable, none more so than the one that controls the tissue type system important in organ transplantation. Just as blood needs to be cross-matched before a transfusion to avoid a fatal immune reaction, so you must match tissue types between donor and recipient when transplanting organs like heart, kidneys or bone marrow. You don't hear of people waiting for a blood transfusion because they can't find a match, but it is a sadly familiar story to hear of patients waiting for months or even years for a suitable heart or kidney donor, often dying before one is found. This is because while there are only four blood groups (A, B, AB and O) there are scores of different tissue types.

I must admit here and now to a serious personal weakness. I have a complete mental block when confronted with the bewildering variety of tissue types. Some of my best friends are cellular immunologists who live, work and breathe tissue types, and the Institute where I work is packed with them. Yet something switches off in my brain when they start describing the various types. All of them begin with the three letters HLA. Then numbers and letters are tacked on to the end: HLA–DRB1, HLA–DPB2, HLA–B27 and so forth. Time and again I go to seminars which kick off with a slide showing a table of this horrendous alphanumeric mélange. For years I concentrated, thinking it would sink in eventually if I tried hard. After all, I have to teach this stuff in my genetics classes. But to no avail. I reluctantly conclude that I am genetically incapable of understanding tissue types beyond knowing that there are an awful lot of them. Which, fortunately, is all you need to know as well. Since there are lots of them, and there are quite a lot of data from Polynesia, South America and south-east Asia, it is relatively easy to track them;

and sure enough, most of the tissue type connections are between Polynesia and Asia. But not all. A type called HLA–Bw48 is very rare everywhere except among Polynesians, Inuit and native North Americans. However, though there is certainly plenty of variation, the evolutionary connection between the different types was not known. So, for example, you couldn't tell whether HLA–Bw48, the type found also in North America, was closely related to other Polynesian types or not. Compare that to the situation with the mitochondrial DNA from Rarotonga. We know that there are three types; we also know that two of them are very closely related to each other, while the third is not. That, as we will see, is an enormous help. We can search other lands not only for the Polynesian mitochondrial types themselves, but for others that are closely related to them as well.

By the time I had planned the return trip, and persuaded the Royal Society to pay for it – after all, they had paid for Cook's first voyage to Tahiti, as I pointed out in my application – data from native North and South Americans produced by other researchers had begun to circulate. Just as there was one cluster in the Rarotongan sample (if we include the two closely related types in a single cluster and forget about the single sequence from the 'tourist'), so there were four main clusters in the Americas. Three of these had quite different mitochondrial DNA sequences; the fourth was rather similar to the main Rarotongan sequence of 189, 217, 247, 261, but with variants at positions 189 and 217 only. This looked very interesting. Moreover, both the native American and Rarotongan DNAs shared another unique feature. At the opposite side of the mitochondrial DNA circle from the control region that we had sequenced, a small piece of DNA, only nine bases long, was missing. This definitely increased the chances that the American and Polynesian types were related. Things were looking up for Heyerdahl.

I had heard that in Hawaii Rebecca Cann, one of the authors with Allan Wilson of the original 1987 paper on mitochondrial DNA and human evolution, was studying the DNA of native Hawaiians. This is hard work because, unlike in Rarotonga, there are very few of them left. Two hundred years of immigration, mainly from Asia and America, have reduced the native Hawaiians to a fringe population, many of them living a marginal existence – an all too familiar legacy of colonialism. However, schemes have recently been introduced by which special grants and scholarships are awarded to those who can prove they are of native Hawaiian ancestry. One way of proving this ancestry is through DNA testing; so there was an extra incentive to find out about the mitochondrial genetics of the native Hawaiians.

On my return visit to Rarotonga I arranged to call on Becky Cann in Hawaii, where we sat down in her lab with her postgraduate student, Koji Lum, to compare results. It didn't take long to discover that we had both found the same major Polynesian type, with the deletion and the same control region variants. This was very exciting, and confirmed the link between the people of Hawaii and those of Rarotonga, three thousand miles to the south. Already I was imagining the enormous ocean distance that separated the two island groups, and the fantastic voyages that must have carried these genes across the sea. Even though it was not unexpected, given the wealth of evidence from the days of Captain Cook onwards that connected all the Polynesians to a common ancestry, just seeing the proof was thrilling. Reluctantly, Becky left to prepare for a seminar, leaving Koji and me in the office sharing our admiration for the voyages of the Polynesians that had carried these genes to Rarotonga and Hawaii.

What followed was one of those rare moments in science when something is revealed that has never been seen before. I was about to pack away my data when I remembered the unusual Rarotongan sequence that I had interpreted as belonging to a tourist and more or less forgotten about. I turned to Koji and asked him if he had ever seen anything like it in native Hawaiians. He agreed to have a look and unpacked his own sheets of results. There was one that stood out from the rest. I laid out my sheet, rather like a roll of wallpaper – this was in the days before laptops – on which the Rarotongan sequences were drawn out, and soon located the unusual sequence. At first Koji's and my sequences looked completely different; then we realized that we had been reading them from opposite ends. I turned mine around, and began to go through the strange Rarotongan sequence. I read from the left-hand side. The first variant was at position 144.

'Do you have anything with 144?' I asked.

'Yes,' said Koji.

I carried on four more bases to 148. 'Anything at 148?'

'Yes, in the same sample,' he replied.

I could feel the thrill of discovery tingling up my spine. I carried on. '223?'

'Yes.'

'241?'

'Yes.'

I accelerated. '293?'

'Yes.'

'362?'

'Yes.'

They were identical. We both looked up at the same time. Our eyes met and two huge, silent smiles shone out from our faces. This was not the DNA of a tourist at all. Discounting the remote possibility that I had accidentally collected a blood sample from a native Hawaiian on holiday in Rarotonga, this had to be a second genuine Polynesian DNA type that had reached into the Pacific as far as the Cook Islands and Hawaii. But where had it come from? It would take another six months to find out.

I flew down to Rarotonga, more determined than ever that we would solve the mystery surrounding the origins of the Polynesians. When I got there, Malcolm, my host from my first visit, arranged for me to meet the man who ran the Prime Minister's office. In most countries this would be quite impossible, but in Rarotonga it was accomplished at Malcolm's Christmas party on the beach. It was fortunate that I met Tere Tangiiti and arranged an appointment early on in the proceedings; because my abiding memory of that party was not of making a crucial diplomatic contact, but of the colour blue: the colour of Curaçao which, mixed with champagne, makes the cocktail Blue Lagoon. Blue Lagoon, seafood omelettes and my digestion don't mix. I was soon to discover the interesting scientific fact that whatever it is they use to colour Curaçao, it is not destroyed in the human stomach. Ten years later I still feel sick at the sight of it.

I needed to get the permission of the cabinet and the cooperation of George Koteka at the health department to collect a substantial DNA sample from Rarotonga and the other islands. I met the cabinet in the Prime Minister's office above the post office, and they could not have been more helpful. Within a few weeks I had collected five hundred samples from Rarotonga, Atiu, Aitutaki, Mangaia, Pukapuka, Rakahangha, Manihiki and even from the tiny atoll of Palmerston (population sixty-six). I packed them carefully in ice and took them back to Oxford.

7

THE GREATEST VOYAGERS

The Institute of Molecular Medicine, where my laboratory is based, is built around the pioneering work of its first director, Professor Sir David Weatherall. His research over the past twenty-five years has been focused on the inherited diseases of the blood, in particular those involving the main component of red blood cells – haemoglobin. These diseases are not particularly common in northern latitudes, but have a quite devastating effect on public health in parts of Africa, Asia and Mediterranean Europe. The main diseases, sickle cell anaemia in Africa south of the Sahara and thalassaemia in Asia and Europe, kill hundreds of thousands of children every year. The causes of all this misery are tiny mutations in the haemoglobin genes, which very slightly alter the oxygen-carrying properties of the red blood cells. In sickle cell anaemia, the usually circular red blood cells visibly change shape, as the name implies, and can no longer slide past each other in the narrowest of blood vessels. This clogs up the flow of blood to vital tissues. In thalassaemia the haemoglobin itself forms clumps inside the red blood cells, which are then destroyed in the spleen. Either way, the anaemias can be fatal if left untreated; and still the only effective remedy is repeated blood transfusions which – quite apart from the side-effects these cause by overloading the body with iron – are beyond the public health budgets in most of the affected regions.

Why do these diseases occur in some places and not in others? The answer is – malaria. Sickle cell anaemia and thalassaemia are found principally in parts of the world where malaria is, or has been, endemic. Both diseases, in order to develop, require a double dose of the mutant haemoglobin gene, one from each parent. Many inherited diseases follow the same pattern; among Europeans the most familiar is cystic fibrosis, where the parents are carriers with one copy each of the mutant gene but no symptoms of the disease. For a reason that even now is not entirely clear, the parasite that causes malaria finds it difficult to infect the red blood cells of sickle cell anaemia and thalassaemia carriers, who as a consequence become at least partially resistant to the disease. Over many generations, this resistance leads to a spread of the haemoglobin mutations in the malarial regions through the forces of natural selection. However, while the mutations are good for carriers, they can be devastating for their children, because some of the offspring of two carrier parents get the double dose of haemoglobin mutants and develop the potentially fatal anaemias. This cruel balance of carrier advantage and offspring elimination keeps the haemoglobin mutants at a high frequency wherever malaria is found. Malaria does not cause these diseases directly, but does so indirectly by allowing, indeed encouraging, the haemoglobin mutations – which are the real cause – to survive and prosper. So, even if you eliminate malaria, you do not at once eliminate these diseases. In Mediterranean Europe – Sardinia, Italy, Greece, Cyprus and Turkey – programmes to eradicate the mosquitoes which carry the malarial parasite have virtually eliminated malaria – but not thalassaemia. Tens of thousands of people still carry the haemoglobin mutations, and only an entirely different programme, built around the genetic testing of prospective parents to see if they are carriers, is reducing the incidence of the disease.

Many people from the Mediterranean have emigrated to different parts of the world, in particular to the United States and Canada, Australia and Britain. With them, literally inside them, travel the thalassaemia genes, so that the disease is also encountered in these communities. For the same reason, forced immigration on slave ships from west Africa introduced the sickle cell gene to North America, where sickle cell anaemia is still encountered, even though there is no malaria. Gradually, over many generations, it will fade from these populations as the mutations are eliminated either by active counselling programmes or simply by the deaths of those who have the disease. Without malaria to help it along it will suffer the ultimate fate of all disease genes – extinction by natural selection.

Unravelling the roots of sickle cell anaemia and

thalassaemia has had a major influence on genetics. It is no exaggeration to say that without the examples of these two diseases to guide researchers, very few of the great advances that have been made since the mid-1980s in finding the causes of genetic diseases would have happened. It was studies of the inherited anaemias that convinced scientists and doctors that simple mutations in genes did indeed cause disease.

The advantages of all this work for me, in my search for the origins of the Polynesians, were far more prosaic. It was field work in the islands of south-east Asia and Oceania, mainly Papua New Guinea, Vanuatu and Indonesia, that finally proved the connection between thalassaemia and malaria. The thalassaemia genes were found only in the low-lying, swampy areas near the coast, where malaria was common, while in the mountainous interiors, where the mosquitoes could not survive the high altitude, the troublesome genes were virtually absent. As a result of this research, the freezers in the Institute of Molecular Medicine were full of DNA samples from these islands. I needed to look no further than the first floor of the Institute where I worked to augment my own samples from Polynesia with a fabulous collection which covered more or less the entire route from south-east Asia into the remote Pacific. If the Polynesians had come that way, surely we would find their mito-chondrial DNA scattered along the route.

Over the summer of 1992 I sequenced over 1,200 mitochondrial DNAs. The first thing to do was to see whether we could find any with the small deletion.

Nineteen out of the twenty Rarotongans were missing this tiny segment and it was very easy to test for it. And there it was: very common in Samoa and Tonga; less common further west in Vanuatu and the coast of New Guinea. The deletion was even less frequent in Borneo and the Philippines, but still there far to the west among the native Taiwanese. This looked like good evidence for an Asian origin. But remember that we knew from published work that the same tiny deletion was also to be found in North and South America. Were we going to find ourselves in the same frustrating situation as everyone else who had tried to use genetics to solve the puzzle, unable to differentiate between a gene that had arrived in Polynesia directly from Asia and one imported indirectly via the land bridge to America? Our only hope was that the control region sequence itself would be able to tell the difference.

The common sequence in Rarotonga, and from the lab in Hawaii, had variants at 189, 217, 247 and 261 as well as the tiny deletion. The other, less frequent but obviously related sequence had variants at 189, 217 and 261 but not 247. As film after film peeled itself out of the developing machine over the next few weeks, I got very good at recognizing the particular pattern of bands that meant we had found the Polynesian sequences. There they were, spread back along the island trail to Polynesia. The further west we went and the closer we got to the Asian mainland, the rarer the full sequence with 247 became, while a new type with just 189 and 217 began to appear, reaching its highest concentration among the Ami, Bunum, Atayal and Paiwan from

Taiwan. The record of the whole amazing journey was there. I rang as many people as I could think of who might have new mitochondrial sequences from native Americans. I had to be sure that 247, the defining variant of Polynesian mitochondrial DNA, was not abundant in the Americas. No-one had seen it. Not even once. Heyerdahl was wrong.

I could not help feeling a tinge of disappointment that I had been unable to vindicate the man who had inspired a generation with his voyage in *Kon-Tiki*. But there it was. His theory had wilted under the fierce spotlight of genetics. The majority opinion had been proved right: the Polynesians had come from Asia and not America. I never got to know what Heyerdahl himself thought about this. I am sure that, at eighty-three, he has better things to do than defend himself against the awesome power of modern genetics. There was a ripple of applause from the anthropology establishment when we published our results; but these academics were already so sure of themselves and convinced by the weight of evidence for an Asian origin that they were not notably excited by this new infor-mation. To agree with the prevailing consensus is unlikely to disturb the peace. To disagree with it, as I was to find out before long in another part of the world altogether, was anything but peaceful.

The genetic trail into the scattered islands of the vast Pacific was now crystal clear. The ancestors of the Polynesians began their epic journey in either coastal China or Taiwan. This is where the highest frequencies of what we can safely assume to be the ancestral mitochondrial DNA sequence of most Polynesians are found today, with variants at 189 and 217 and the small deletion. We also found in the samples from Taiwan other sequences with extra variants on top of the core 189, 217 pattern but at positions we didn't find in other parts of the region. These are the mutations that have happened in Taiwan since the ancestors of the Polynesians left. By counting up the mutations and multiplying by the mutation rate we can estimate the length of time since the ancestral sequence itself first arrived in Taiwan. As we shall see when we come to explore the genetic landscape in Europe, this is a controversial area in contemporary research. None the less, it was pretty clear from the great diversity of variation on the basic theme of 189, 217 in Taiwan that the sequences had been there a very long time indeed, probably as long as twenty to thirty thousand years.

There are many archaeological signals of a very sudden population expansion in the islands of southeast Asia around three to four thousand years ago, defined by a range of artefacts associated with an agricultural economy. The most significant of these is pottery of a distinctive style called Lapita, with a red surface glaze and tooth-like decorations stamped into the clay in horizontal lines. For archaeologists, pottery with an identifiable style is a real bonus. It survives for thousands of years in the ground, and a similarity of ceramic style can connect settlements that are geographically far apart. It doesn't automatically mean that the people who used the pottery were biologically related, but it is a certain sign of contact between the different places. Within a period of only five hundred years, beginning three and a half thousand years ago, Lapita settlements appeared on the coast of many of the islands in the western Pacific, from the Admiralty Islands north of New Guinea to Samoa in western Polynesia. Supporters of the Asian origins of Polynesians had always associated this rapid expansion, which implied a sophisticated seagoing capacity, with the people who ultimately colonized the whole of Polynesia. The mysterious absence of Lapita pottery on the islands to the east of Samoa was explained by the lack of a suitable clay. Now that the genetics had come down firmly in favour of an Asian rather than an American origin for the Polynesians, could we say anything new about where this remarkable expansion of people and pottery might have begun?

First of all, the complete absence of the variant at position 247 in Taiwan made it extremely unlikely that it had started there. If it had, then I would have seen plenty of variant 247 in Taiwan. In fact, I never see variant 247 west of Borneo. So the rapid Lapita expansion is only supported by the genetics if it began somewhere to the east of Borneo. I did see 247 in the Moluccas, an island group in Indonesia, and it has been there long enough to accumulate additional mutations. My best estimate for the place of origin of the remarkable Lapita Polynesians would be somewhere in that island group. From there, the mitochondrial trail leads out into the Pacific, to Hawaii in the north, to Rapanui (Easter Island) in the east and to Aotearoa (New Zealand) in the far south.

All this is clear from the main Polynesian type. But what of that strange, rare sequence that I had found in the blood from a single outpatient in Avarua hospital and Koji Lum had found in one native Hawaiian? Could this be the faint echo of Heyerdahl's American Polynesians? We had certainly found this sequence all over Polynesia after our extensive sampling, though it was never common; but none of my contacts had seen anything like it in North or South America. Then we found a single example in Vanuatu and two more from the north coast of Papua New Guinea. However, only when I tracked down some old samples from the mountainous interior of New Guinea did I find this sequence in abundance. This was mitochondrial DNA that had been handed down to the present-day inhabitants from the earliest settlers of that huge island – settlers who, according to the dating of early archaeological sites, had made their way there at least forty thousand years ago in the same ancient migration that had carried the first Australians to that vast continent. So the direct maternal ancestors of the mysterious outpatient from Avarua hospital had spent almost forty thousand years on the island of New Guinea before joining a Lapita voyaging canoe heading east into the unknown.

From the north coast of New Guinea a line of islands, each visible from the previous one, stretches out into the Pacific as far as the Solomon Islands. These are high islands with mountain peaks which can be seen on the horizon either before setting out or, at the very least, before losing sight of your departure point. This comparatively safe navigational technique had already taken the earliest settlers of New Guinea up past New Britain and New Ireland and down the main chain of the Solomon Islands as far as San Cristobal thirty thousand years ago. But this was the end of the pier. Beyond that was the open sea with the nearest land, the islands of Santa Cruz, still three hundred kilometres away far beyond the horizon. There is no archaeological evidence of any settlement beyond the Solomons until the arrival of the Lapita people twenty-seven thousand years later.

Two crucial developments enabled the new wave of colonists to launch into the unknown. The first was the development of the double-hulled voyaging canoe. These magnificent vessels reached enormous sizes. The first Europeans to reach Polynesia saw canoes over 30 metres long, and smaller versions are still used today. The double hull prevents capsizing in the same way as the outrigger on a catamaran. The vessels had a prow at each end, and so could be tacked across the wind and then reversed without turning round. These were the vessels that carried the Polynesians into the Pacific; the complementary and equally crucial development was a highly sophisticated set of navigational skills. Whereas the earliest settlers had managed to reach Australia, New Guinea and the Solomon Islands by steering to visible targets, the Polynesians sailed off into a void, not only unable to see land but not knowing if there was any. Their progress can be followed through the dating of archaeological sites. They quite quickly settled Santa Cruz and the islands of Vanuatu, paused before the 750 kilometre crossing to Fiji and beyond to Samoa and Tonga, then paused again before pushing on to the limits of Polynesia. They reached Easter Island and Hawaii about fifteen hundred years ago and, last of all, New Zealand about twelve hundred years ago. They had reached every island in this vast ocean in a little over two thousand years. How did they do it?

Well supplied with food and water, the canoes set off against the prevailing wind. This might seem like hard work, but it did at least ensure a relatively safe return journey, sailing back downwind to the home port, navigating by the stars. It is fairly straightforward to travel along a line of latitude by keeping a rising or setting star in the same position relative to the canoe each night. When it is time to return home, it is a simple matter of reversing direction and sailing down-wind guided by the same stars – simple in theory, but in practice still fraught with danger. It must have been all too easy to miss the home island, especially if it was passed in the night or in a storm. There must have been many losses.

More remarkable still are the signs the Polynesians used to detect the presence of unseen land. Cloud formations above high islands betray their presence over the horizon. The opalescent blue-green of low-lying atolls is reflected in the clouds under the right conditions. The outward and homeward flight directions of birds known to nest on land give clues. Floating debris shows there is land upwind. These are the visible signs. But the early voyagers not only saw their way ahead; they felt it too. Even now traditional navigators can detect the change in the way the sea moves. The underlying swell sweeps across the ocean, but is reflected back from the islands – just as ripples from a stone tossed into a pond bounce back from the edge. Even a hundred miles from land an experienced navigator will use his feet to feel for the interference pattern as one swell crosses another.

That is *how* the Polynesians arrived. It is not *why*. What drove them to these quite extraordinary feats of exploration is still a mystery. It cannot be that they needed to keep moving on to satisfy the economic needs of a growing population. Many of the islands en route to the remote Pacific are large and fertile, and are not densely populated even today. Perhaps it was an insatiable urge to explore the unknown. They must certainly have voyaged right across the Pacific to reach South America. We know this from the evidence of the *kumara*, the sweet potato, which was and is cultivated all over Polynesia. There is no dispute that the sweet potato is a true Andean crop, so this has to mean there must have been at least some contact between the native South Americans and the Polynesians. The genetics rules out Thor Heyerdahl's explanation that the sweet potato was brought into the Pacific by the people who he thought had colonized Polynesia from South America: we had proved beyond any doubt that the colonization of the Pacific had happened in the opposite direction. For the sweet potato to have been imported from South America, the Polynesians must have found their way right across the Pacific. But they left no visible genetic trace in South America. To my

knowledge, not a single sample of Polynesian mitochondrial DNA has ever been found there. However, I
did eventually find two mitochondrial sequences from
Tahiti in French Polynesia that matched published
sequences from Chile. I like to think these may have
been the faint echo of women who had joined
the Polynesians on their return journey back into the
Pacific from South America, but I cannot prove it.

The Polynesians discovered and settled Aotearoa, the
'land of the long white cloud', which we know as New
Zealand. The genetics proves that, too, beyond any
doubt: the Maori of Aotearoa share exactly the same
mitochondrial DNA as their cousins in Polynesia. This
agrees perfectly with Maori oral tradition, which relates
that a fleet of eight to ten voyaging canoes set out from
central Polynesia, perhaps even from Rarotonga itself,
and eventually reached Aotearoa. They found a strange
but fertile land with no human inhabitants but full of
creatures they had never before encountered, including
the huge flightless moa, related to the ostrich (and
eventually hunted to extinction).

Travelling so far to the south, the voyagers would
have left themselves little hope of return had they not
discovered land. Getting to Aotearoa was not a simple
matter of sailing into the wind along a line of latitude
in the knowledge that if no land was encountered it
was only necessary to reverse direction and travel home
along the same line with the wind behind them. The
voyage to Aotearoa took them across latitudes and
far to the south of the reliable trade winds into a part of
the ocean where the winds were far less predictable.

This was another level of maritime exploration altogether, of such maturity and daring that I feel sure the Polynesians must also have reached the coast of Australia, so accomplished were they in the ways of the ocean. But if they did land here, they left no trace. Perhaps they only felt comfortable settling uninhabited lands. Did they, I wonder, sail south of Australia and right across the Indian Ocean to Madagascar – unpopulated then, and now at least partially inhabited by people speaking a similar language? Could they have done that? I'm quite sure they could. Did they? One day, the genes will tell us.

I still feel a sense of excitement when I think about the work in Polynesia. I had found myself on an island in the middle of the Pacific staring out at the ocean beyond the reef and overcome by a burning desire to find the answer to a question. I really wanted to know where the Polynesians had come from. It was a quest born out of pure curiosity; and it had delivered the answer, clear-cut and unequivocal, in a little over three years. Having seen how decisively mitochondrial DNA had settled the matter of Polynesian origins, I was very confident of its inherent ability to solve even harder questions in an arena much nearer home.

8

THE FIRST EUROPEANS

The ten-year excavation at Boxgrove near the cathedral city of Chichester in Sussex, England had been productive but not spectacular. Boxgrove is a quarry now; but half a million years ago it was a narrow coastal plain between chalk cliffs and the open sea. The sand and gravel that are now quarried were washed in there by later floods created by the catastrophic melting of later Ice Ages. Over several years Boxgrove had yielded up flint tools and animal bones with cut marks that showed that the carcasses had been deliberately butchered using the razor-sharp stones. If you doubt how sharp these could be, then try knocking a flake from a large piece of flint. It is quite sufficient for a close dry shave. The shaped stones and the bones were obvious signs of human occupation – but there had been no sign of the humans themselves. English Heritage, the government agency that had been paying for the excavation, had already made it clear that it would not fund any more fieldwork. At the beginning of November 1993, with only weeks to go before the excavations were finally abandoned, the archaeologists in charge marked out one last trench and Roger Pedersen started digging.

Roger, one of the army of devoted volunteers that are the lifeblood of all archaeological digs, proceeded to make a start on the trench with his trowel. After two weeks he had dug down into the layers of sand, recording the orientation of every artefact he came across. It was slow and painstaking work, not made any easier by the cold, wind and rain. Then, just after lunch on Friday, 13 November 1993, he found a bone – the shin-bone of a very early human. He had uncovered a fragment of Boxgrove Man. And he had saved the dig.

I was shown the bone soon afterwards, and though I am no expert, even I could see how very thick the walls surrounding the central marrow space were, compared to a modern bone. This was the shin-bone of a massive, heavily built human. But was it the bone of an ancestor? This straightforward question goes to the very heart of the controversy over human origins, for one simple reason: whereas every living human (or animal or bird, for that matter) has ancestors, it does not automatically follow that every human fossil had descendants. Boxgrove Man might be an ancestor of twenty-first-century modern humans, or he might belong to a species that is now extinct.

Exactly the same argument surrounds every human fossil. There are many sites of great antiquity in Europe, in Asia and especially in Africa that have yielded what we would have little trouble recognizing as signs of human activity. These are mainly the remnants of shaped stone tools, which obviously survive extremely well. Occasionally, as at Boxgrove, there are animal bones with deliberate cut marks. And very occasionally, there are actual human bones. These fantastically rare and celebrated specimens have been studied and argued over by palaeontologists for decades. Their names – *Homo habilis*, *Homo erectus*, *Homo heidelbergensis*, *Homo neanderthalensis* – reflect the to and fro of attempts to pigeon-hole them into different species. However, these are species defined on the basis of the anatomical features preserved in the skeletons, particularly the skulls, and not in the biological sense of different, genetically isolated, species who are incapable of breeding with any other. It is an operational classification with no evolutionary consequences. From the shapes of the bones alone there is simply no way of knowing whether humans (I use the term 'human' to include everything in the genus *Homo*) from different parts of the world were capable of successful interbreeding. If they could interbreed, then this opens up the possibility of their exchanging genes and spreading mutations around. They would all be in the same gene pool. But once the different types of human become incapable of interbreeding, they can no longer exchange genes. They become different *biological* species with isolated gene pools. Their evolutionary pathways are irreversibly separated, setting off in different directions with no turning back. If two or more of these species later come into conflict for space or resources, then, unless a compromise is reached, one species will become extinct.

It is this question that lies behind one of the

longest-running and most deep-seated controversies in human evolution. Are the different species defined by palaeontologists – *Homo erectus*, *Homo neanderthalensis* and ourselves, *Homo sapiens* – all part of the same gene pool or not? Or, to put it another way, are modern humans directly descended from the fossils found in their part of the world, or are many of these the remains of now extinct genetically separate human species?

There is no serious doubt that all humans alive today are members of the same species, *Homo sapiens*. Historical events over the last few hundred years have intermingled people from very different parts of the world, producing abundant evidence of successful inter-breeding between all possible combinations. At least, I say that without being completely sure that the opportunity has arisen for absolutely *all* possible combinations to have been tried; but I'm certain that there would be no genetic barrier to success if they were.

The human fossil record, incomplete and patchy though it is, consistently points to Africa as the ultimate origin of all humans. In Africa and only in Africa is there a sensible progression of fossils covering the past three million years and showing intermediate forms from ape to man. Judging by the fossil record, early humans spent at least another million years in Africa before beginning to venture further afield. Remains in Java and China resemble much older *Homo erectus* fossils from Africa not only in their general overall physical appearance but also in the types of stone tools found at the sites. *Homo erectus* was certainly convincingly human, fully upright with a large brain and capable of making and using sophisticated stone tools. But there are no signs of any more primitive, intermediate fossils anywhere outside of Africa. However, while the fossil record is unambiguous in identifying Africa as the cradle of humanity – a conclusion with which very few would nowadays disagree – we should bear in mind some of its limitations. No human fossils have ever been found in west Africa, for example. That does not mean humans were not there until recently; only that the tropical rainforests are not good places to turn into a fossil when you die. No fossils of any of the great apes – gorillas, chimpanzees or orang-utans – have ever been found. As far as the fossil record is concerned, they never existed; and yet we know from the evidence of our own eyes that they did, and do.

Though the fragments of Boxgrove Man and a handful of others are the only glimpses we have of the very earliest European humans, who lived over half a million years ago, the more recent history of Europe is inextricably linked to one dominant form – the Neanderthals. In 1856, workmen quarrying limestone in the Neander valley near Düsseldorf in Germany had just blasted out a small cave and were cleaning away the debris when they came across part of a skull, then thigh-bones, ribs, arm- and shoulder-bones. They thought at first they had come across the remains of an extinct cave bear, an almost routine find in that part of Europe. Only by chance did they happen to mention their discovery to a local schoolteacher and enthusiastic naturalist, Johann Karl Fuhlrott, who realized as soon as he saw the remains that this was no cave bear. Exactly what it was remained controversial for several years. The skull was not that of an ape; but then, with its massive brow ridges, it wasn't exactly human either. For a start, how old was it?

The Neander valley – in German, Neanderthal – bones were found at a time when the biblical account of the creation was coming under attack from geologists who could not accept that the world was only a few thousand years old. Three years later Charles Darwin published *On the Origin of Species* and the status of the Genesis story as literal truth started to crumble. Gradually the idea that humans had really ancient predecessors became more widely accepted; and it looked increasingly as though the Neanderthal 'man' might be one of them. But this conclusion was reached only after discounting the usual crop of red herrings that accompanies unexpected finds like this. They ranged from the sublime – this was the skull of a man with a mystery bone disease which caused the thickening and the brow ridges – to the ridiculous – it was the skeleton of a Cossack horseman who had been injured in the Napoleonic Wars and crawled into the cave to die. Without his sword and uniform . . . ?

Over the next hundred years, several other fossils were found that conformed to the same pattern: heavy build; large braincase (actually slightly larger than the modern average), presumably to accommodate a large brain; no real chin; a prominent nose; and those distinct, heavy brow ridges. The fossils came from

Gibraltar and southern Spain – in fact, the first Gibraltar specimen had been excavated in 1848, eight years before the discovery at Neanderthal, but was neglected. They were found in Belgium, France, Croatia; and from further afield in Israel, Iraq and as far east as Uzbekistan. The stone tools found at Neanderthal sites were more advanced than those associated with their predecessors, though not a great deal. They may have intentionally buried their dead and even cared for the sick and dying. These were not the unreconstructed brutes of popular imagination. But still the question remained: were these people the ancestors of modern Europeans or just another evolutionary dead end?

The same question applies to other parts of the world. Are modern Chinese the descendants of the people whose million-year-old remains were found at Zhoukoudian near Beijing? Did the ancient people of Ngandong in Java eventually become the modern native Australians and Papuans? That is certainly the view of an influential and vocal school of contemporary physical anthropologists – the multi-regionalists. They see the change in human physical characteristics over the past million years from the robust, heavy-boned ancestors to their slender (at least theoretically!) and light-boned descendants as a gradual process of adaptation happening at different speeds in different parts of the world. Though geographically remote from each other, there has been enough contact among these groups to maintain a common gene pool and allow modern *Homo sapiens* to breed successfully with whomever he or she wishes. Always assuming they get the chance.

The opposite camp – the replacement school – fiercely contests this view of continuity. Their view is that both the Neanderthal and the Zhoukoudian and Ngandong fossils, also known as Peking and Java man, are remnants of extinct human species that were replaced by a much more recent expansion of *Homo sapiens* out of Africa. The fossil evidence put forward in support of this contention is the abrupt appearance in Europe about forty-five thousand years ago of humans with much lighter skeletons and skulls which are virtually indistinguishable from those of modern Europeans. There is no debate, even among the most argumentative of palaeontologists, that these are the remains of our own species, *Homo sapiens*. In Europe, these early examples are known as the Cro-Magnons, named (in the same tradition as the Neanderthals) after the cave site of Cro-Magnon in France: one of the places where, in 1868, such bones were first found. It is inconceivable, according to the replacement school, that a mutation of such magnitude could have occurred as to transform the heavily built Neanderthals into the thoroughly modern-looking Cro-Magnon more or less (in evolutionary time) overnight. The archaeological as opposed to the fossil evidence for an abrupt replacement of Neanderthals by Cro-Magnons is the use of a much more highly developed and delicately crafted set of tools, with flint slivers used for knives, scrapers and engravers; the appearance for the very first time of animal bone and antler as an industrial material; and one more crucial ingredient – art.

The Cro-Magnons had invented representational art. Over two hundred caves in France and northern Spain are adorned with their strangely beautiful and vigorous images of wild animals. Deer and horses, mammoths and bison decorate the walls of the deepest caverns, far from the light of day. These are not crude or child-like drawings but the expression of a mature and accomplished imagery, an abstracted and mystical representation of their world.

Is it possible that the Neanderthals had not only transformed their physical appearance and their technology, but had also become artists as well? The multi-regionalists think just that, and, indeed, see in some remains and stone tools evidence of the intermediate forms you would expect from a gradual transition. But there are no precedents for the cave art anywhere in the lands where Neanderthals have been found. The school of sudden replacement traces the modern anatomy and the improved technology back to Africa, to sites like Omo-Kibish in Ethiopia, which are well over a hundred thousand years old. Even so, although anatomically modern skulls have been found along the trail to Europe in the Near East, principally at Qafzeh and Skhul in Israel, there was no sign there of art.

Without new evidence from a completely different and independent source, genetics, the debate about whether native Europeans were descended from Neanderthals or from the apparently distinct later arrivals, the Cro-Magnons, would have rumbled on unresolved. In all fields of human endeavour where there is a shortage of objective evidence, opinions and people inevitably become polarized into rival camps. Once entrenched, the occupants will not be dislodged; they would rather die than change their minds. Such was the situation when we set out to apply our powerful genetic tools to the conundrum; so we knew the path ahead would likely lead us into a minefield.

9

THE LAST OF THE NEANDERTHALS

Genetics is at its most powerful when it comes to distinguishing between rival theories. In the Pacific it had come down decisively on the side of an Asian origin for the Polynesians at the expense of Thor Heyerdahl's American alternative. Could it do the same for Europe? Could genetics give an equally clear answer to the true fate of the Neanderthals? Were these strange humans a staging post on the way to fully modern Europeans, or were they an essentially different species that was replaced by the lighter-boned, technologically advanced and artistic new arrivals from Africa? This was the principal question I now set out to answer with mitochondrial DNA. Just as the success with the Syrian hamster had given me the confidence in the reliability of the DNA segment known as the control region, so its brilliant performance in the Pacific meant I now felt ready to disentangle the far greater complexities of Europe.

I had discovered the true origins of the Polynesians by studying the genetic variety we found in their modern descendants. The great majority had DNA signatures that were either identical, or very similar to one another. Along the whole 500 base DNA segment that we had routinely sequenced, there was a difference of only one, or at most two, mutations among them. On an evolutionary timescale, these people had all shared a common ancestor very recently indeed. The genetic trail of identical and near-identical sequences led back from island to island, to Taiwan and south China. This is a beautifully laid out record of the incredible voyages of the first Polynesians, easily read in the genes of the modern population. But there are a few Polynesians, around 4 per cent, whose DNA tells a different story. They are closely related to one another within a cluster of sequences but, on average, thirteen mutations distant from the main Polynesian sequences. This cluster did not come from mainland Asia but – as described in chapter 7 – could be traced back to the coast of New Guinea, from where they, or maybe just she, boarded a Lapita canoe and headed east into the Pacific.

The mitochondrial DNA had shown very clearly that the maternal ancestors of modern Polynesians came from two different places – from two very different peoples, who had since become mixed. Would the Europeans also show a clearly mixed genetic ancestry with, perhaps, a 'Neanderthal' cluster and a 'Cro-Magnon' cluster to be found among the modern population? Even though the mixing of Neanderthal and Cro-Magnon genes could have been going on for forty or fifty thousand years, compared to only three or four thousand years in the Pacific, I felt sure that I would still be able to pick out any distinct clusters in

Europe, just as I had in Polynesia. That I felt so confident was entirely thanks to the special inheritance pattern of mitochondrial DNA. Unlike the chromosomes of the nucleus, mitochondrial DNA is not shuffled at each generation. The only changes are brought about by mutation, and forty thousand years is not so very long in mutational time. If there had been substantial interbreeding between the Neanderthals and the Cro-Magnons, we would find the evidence in the modern population.

There was only one way to find out: my research team had to start testing, and on a wide scale. What was going to be the best way of going about it? Whom would we ask, and how? And what would we ask for – a blood sample? There were plenty of questions to be resolved, but one thing I was sure of. If at all possible, we would collect the samples ourselves, rather than relying on older collections. The scientific reason for this was that I wanted to be sure we knew that if a sample came from, say, north Wales, it was from someone whose ancestors came from the same area. We sat down to plan our campaign. Martin Richards, now the senior scientist on the team, thought of approaching local family history societies; but I wasn't sure this route would give us wide enough coverage in a short enough time. Our research grant only had another year to run, and we would need to build a persuasive case, built on results, to gain continued funding for the project. I favoured touring sheep and cattle markets, reasoning that farmers were likely to be the most settled population with local roots going back a long way. But it was Kate Smalley, the third member of the team, who came up with the solution.

Kate had been a teacher before coming into research, and she thought that if we wrote to schools who taught biology in the sixth form we could combine a presentation on modern genetics with a sample collection. This idea had a lot going for it. Kate thought we would get a high uptake if we contacted schools with this suggestion, not just because genetics was beginning to feature more and more in the examination syllabus, but also because it gave the teachers themselves a double period off. She was absolutely right and we had a 100% favourable response from the schools we approached.

Where were we to start? We would need to home in on areas where we could be sure of finding a high proportion of long-established local families. I had been reading some old papers written in the 1950s about blood groups in Wales. One anecdote in particular caught my eye. It was an account of the odd head shapes allegedly found in mid-Wales. Those were the days, thankfully long gone, when skull measurements were a respected source of information for physical anthropologists intent on classifying the whole of humanity into different racial types. According to this account, the heads of some people in mid-Wales bore a close resemblance to those of 'Stone-Age Man', whatever that was. Apparently a hat shop in the market town of Llandysul, not far from Cardigan, regularly had to supply hats made to measure because so many of their customers were unable to fit into the standard sizes. This is not the sort of thing to take too seriously;

but neither should it be totally dismissed out of hand. After all, it was skull measurements that initially led Arthur Mourant to turn his attention to the Basques when looking for the descendants of Europe's 'original' population. So Wales looked like a good place to start, and within a month Kate had organized a week-long tour of the whole principality.

In the early spring of 1992 we set off in two cars, having mapped out a complicated pincer movement whereby two pairs (we were joined by Catherine Irven, who took a week off from another project) would take different routes round the country, meeting up halfway to see how the other was getting on. My car at the time was a thirty-year-old Mk II Jaguar/Daimler that I had bought in a moment of utter madness from a garage forecourt in New Zealand the previous year and had had shipped over. It had a tendency to dislodge its water hoses every so often, causing the coolant to flood out and sending the engine temperature rocketing skywards before eventually conking out. So as well as all the blood sampling equipment on board, I was forced to carry a full toolkit – which was just as well. As we swept into the school in Bala, in central north Wales, there was a loud bang and a foul smell of burning oil filled the car. We pulled up in the car park at one side of the playground and, with the children watching from the classroom windows, I looked under the bonnet to see what had happened this time. There was black oil all over the place and clouds of acrid grey smoke were billowing up from where the oil had hit the exhaust pipes. This wasn't the best way to arrive. I couldn't tackle the job without getting covered in oil; hardly the best way to appear if you want to take blood samples. I shut the bonnet and walked into the school.

Sometimes the problems didn't stop outside. We had let the schools know that we would be happy for them to tell their local papers that we were coming if they wanted to. This had seemed a good idea – until I got to Ysgol-y-Gader in Dolgellau. Sitting with the head teacher, Catherine James, in her office was a reporter from the *Caernarvon and Denbigh Herald*.

'So you are here to do blood tests on the children?' he asked, opening the interview innocently enough.

'Well, yes,' I replied. 'But only as a source of DNA, the genetic material.'

'Why have you come to Dolgellau?' he asked.

I gave a short description of the background to our project and what we wanted to do. I explained that, because of their settled population over the last few centuries, we were particularly interested in the areas of Wales, like Dolgellau, where the Welsh language was still spoken. He didn't look as if he believed me.

'You're really here because of the power station, aren't you?' He looked right at me. 'You want to test the children for mutations, don't you?'

I was stunned. Dolgellau is just twelve miles south of the Trawsfynnyd nuclear reactor. A few months before, news reports had linked mutations found in children living near the nuclear reprocessing plant at Sellafield in Cumbria with their fathers working at the plant.

The expression on the head teacher's face rapidly changed from one of mild interest to intense suspicion. Was her school, was she, being used by undercover agents for the nuclear power industry posing as academics engaged on an innocent-sounding study of Celtic genes?

'Of course not,' I stammered and embarked on a stream of denials and reassurances. I repeated an account of the scientific background, a description of mitochondrial DNA, a summary of our work on ancient bone, and finally what I thought would be an irrefutable certificate of our integrity: 'Anyway,' I said confidently, 'I've just come back from doing the same research in the South Pacific.' That would do it. Or so I thought.

'But isn't that where they test the atom bombs?' he replied, quick as a flash.

I groaned, took a deep breath, and launched myself on another twenty minutes of explanation. Eventually they were both persuaded of our innocence and we could get on.

At the end of my talk to the sixth form, the time came to ask for the blood samples. This was the point where I anticipated some further difficulty. Taking DNA from older schoolchildren (they had to be over sixteen to be able to give legal consent) ruled out taking a large blood sample, and we settled on a drop of blood taken from a finger-prick. This did involve some slight discomfort, and we were worried that no-one would want to do it. At first, to demonstrate how painless it was, I pricked my own finger and dabbed the little drop of blood on to a special absorbent card. Next, the teacher tried it; and one by one the pupils followed. For youngsters who haven't done it before, it does require a little bit of courage. What happened next was an unexpected bonus. Precisely *because* they had done something brave, as soon as they had finished the children shot out of the classroom and round the school – it was the lunch break by then – daring their friends to do the same. A line of suppliants appeared, all swearing they were over sixteen, begging to be sampled not so much because of their intense interest in the project but because they wanted to prove themselves just as courageous as their friends. This wave of bravado spread to the staff room and the kitchens, so that by the start of afternoon classes we had blood samples from all the children old enough to take part, the teachers, the janitors and the dinner ladies.

By the end of the week we had over six hundred blood samples, dried on to cards, from all over Wales – a remarkable haul that far exceeded our expectations. Even though it may not sound a lot, and is only a tiny proportion of the total Welsh population of almost three million, six hundred mitochondrial DNA sequences would be more than enough to get a good idea of the general genetic structure of the principality. Back in the lab we punched out the circles of dried blood from the cards and set about extracting the tiny amount of DNA they contained. Though there are a lot of cells in blood, most of them were no use to us. The red corpuscles, which carry the oxygen and make blood red, are so specialized that they do not need a nucleus or mitochondria; so these superfluous components are evicted early on in the life of the cells, which consequently do not have any DNA in them. Only the white blood cells, whose job it is to seek out and destroy invading bacteria and viruses, retain their own nuclear and mitochondrial DNA. White blood cells make up only 0.1 per cent of the cells in blood, so that while a drop of blood might have fifty million cells in it, only fifty thousand of them contain any DNA. But this is still plenty for the exquisitely sensitive DNA amplification method to work on. We followed the same recipe for getting DNA out of the blood spots as forensic laboratories use on blood-stained clothing prior to taking a genetic fingerprint. This involved boiling the dried blood spots in dilute alkali, which splits the cells open and dissolves the DNA, then adding a resin to absorb the iron which has been leached out from the red blood cells and which would otherwise interfere with the DNA amplification reaction. It worked very well indeed, and before long we had our first 100 Welsh mitochondrial DNA sequences.

Compared to the relative simplicity of the Polynesian sequences, the Welsh results were all over the place. There was no sign of a clear-cut distinction in Wales analogous to what we saw in Polynesia, where the two separate clusters were so clearly the result of a mixture of people from very different origins. It looked as if we had a small number of little clusters which were all quite closely related to each other, rather than two big clusters separated by a large number of mutations. This did not look like the mixture of two very different types of mitochondrial DNA that we would have expected if the people of Wales had a mixed Neanderthal and Cro-Magnon ancestry. If Wales was going to be representative of Europe as a whole, then we were looking at a comparatively recent shared ancestry for everybody.

Along the 500 base segment of the mitochondrial DNA control region, the average distance between any two people among the volunteers from Wales was three mutations. Remembering the rate at which the mitochondrial DNA clock ticks, so that two people with a single mutation between them can be said to have shared a common maternal ancestor about ten thousand years ago, the result from Wales showed that the average length of time it was necessary to go back in the past to connect any two people from Wales was only thirty thousand years; and even the most extreme difference between two of our volunteers, which was eight mutations, meant that they shared a common ancestor only about eighty thousand years ago. Although this is an enormously long time, it is nowhere near long enough for one of them to have been the descendant of a Neanderthal and the other of a Cro-Magnon. Unless the palaeontologists of the replacement school were way off the mark, Neanderthals and Cro-Magnons last shared a common ancestor at least two hundred and fifty thousand years ago. That means that the mitochondrial DNA of a Neanderthal descendant and that of a Cro-Magnon descendant would differ, on average, by at least twenty-five mutations. The biggest difference we saw in Wales was only eight. This was not a mixed population of ancient and modern humans. Either the Welsh were all Neanderthal or they were all Cro-Magnon. But which?

The few sequences coming in from other parts of western Europe did not suggest to us that the Welsh were completely different from the rest. The stark alternatives of 100 per cent Neanderthal or 100 per cent Cro-Magnon ancestry seemed to apply throughout Europe. The acid test to distinguish which of the two competing ancestries was the real one would be a comparison between the European sequences and the corresponding data available from other parts of the world, which included our data from Polynesia. If there were big differences, of the order of twenty-five mutations or more, between the Europeans and the Polynesians, then the votes would go to a Neanderthal ancestry for all modern Europeans. If the differences were far less than that, it would mean a 100 per cent Cro-Magnon ancestry for Europeans, and a victory for the replacement school at the expense of the multi-regionalists.

When we looked at the data, the biggest number of mutations we found between two people was the fourteen that separated Teri Tupuaki, a fisherman from Mangaia in the Cook Islands, and Mrs Gwyneth Roberts, who cooks the school lunches in Bala, north Wales. These two people, half a world apart, between them solved a puzzle that had divided scholarship for most of the twentieth century. Europeans were not that much different from the rest of the world; certainly nowhere near different enough to justify believing that they were all descended from Neanderthals. And since it was all or nothing, the Neanderthals must have become extinct. All modern Europeans must today trace their ancestry back to much more recent arrivals – to the Cro-Magnons, with their lighter skeleton, their much improved flint technology and their wonderful art. This was an absolute replacement of one human species by another. Whether it was an active and violent process, with the newcomers, our own ancestors, evicting or even killing the resident Neanderthals, or whether it was their technological and mental superiority that gradually marginalized the older inhabitants, the genetics alone cannot say. It is clear from the fossil record that the Neanderthals hung on for at least fifteen thousand years after the first Cro-Magnons reached western Europe some forty to fifty thousand years ago. When the last Neanderthal expired – probably in southern Spain, where the most recent skeletons have been found – his or her death drew a line under another phase in the human occupation of Europe. An era that had lasted for a quarter of a million years ended, finally and irreversibly, in a cave in southern Spain about twenty-eight thousand years ago.

I confess to some surprise, and some disappointment, that the replacement was so complete. Even though we have now sequenced the mitochondrial DNA of more than six thousand Europeans, we have never yet found a single one that is even remotely credible as a Neanderthal survivor. We certainly haven't sequenced everybody, nor have we had a chance to receive samples from every corner of the continent. I retain the hope that one day, when I look at a batch of read-outs from the sequencing machine, I will find a sequence so different from the rest that it calls out as the faint echo of a meeting between Cro-Magnon and Neanderthal which led to the birth of a child. If we ever did find one, we could not miss it. In 1997, DNA was sequenced from the very first Neanderthal skeleton from the original find in the Neander valley. It had twenty-six differences from the average modern European, more or less exactly as predicted for a species that last shared a common ancestor with *Homo sapiens* a quarter of a million years ago. The DNA sequence of a second Neanderthal, this time from the Caucasus mountains, was reported in the scientific literature in 2000. It was equally different from modern humans. These were not our ancestors.

In 1998, the partial skeleton of a child with anatomical features intermediate between Neanderthal and Cro-Magnon was found in Portugal. Could this be evidence of interbreeding between the two types of humans? Perhaps. The child's DNA has yet to be tested. But if this interbreeding were a frequent occurrence, then surely we would see the evidence in the modern mitochondrial gene pool, and we just don't. If the interaction between Neanderthal and Cro-Magnon resembled more recent historical encounters between new arrivals and the original inhabitants of a territory, then we might expect the matings to be between Cro-Magnon males and Neanderthal females rather than the other way around. In that case, mitochondrial DNA would be an excellent reporter of these encounters, since while the offspring would have an equal mixture of nuclear DNA from both parents, their mitochondrial DNA, inherited from their mother, would be 100 per cent Neanderthal. As a geneticist it is very hard for me to imagine that social and other taboos were so strong that this never happened; but we must continually return to the evidence and the complete absence of any Neanderthal mitochondrial DNA in modern Europe.

Could it be that the matings did occur but did not produce viable and fertile offspring? There are many examples from the animal world of hybridization between different species leading to perfectly healthy yet sterile offspring. The textbook example is the mule, the fruit of accidental or intentional matings between a male donkey and a female horse. The horse and donkey genes must be mutually compatible because mules are strong, healthy and fully functional, except when they come to breed. That's because donkeys and horses have different numbers of chromosomes. Horses have 64 chromosomes, donkeys have 62. All mammals, including humans, inherit a half set of chromosomes from each parent to make up their full complement. So a mule gets 32 chromosomes from its horse mother and 31 from its donkey father – and so ends up with 63 chromosomes. That is not a problem for the body cells of the mule, because both horse parent and donkey parent genes can be read irrespective of which chromosome they are on. It's only when the mule tries to breed that the confusion starts. For one thing, being an odd number, it is impossible to get a half set from 63 chromosomes. For another, the scrambling of the chromosomes that occurs at each generation leads to mule sperm and mule eggs with two copies of some genes and none of the others. For both these reasons, mules cannot produce offspring.

Were the encounters between Neanderthal and Cro-Magnon also doomed to produce only one generation of infertile hybrids because they had different numbers of chromosomes? Our nearest primate relatives, the great apes (gorillas, chimpanzees and orang-utans) have one more chromosome than we do. At some point in the six million years since humans and great apes split away from our mutual common ancestor, two chromosomes that are still separate in the great apes fused together in the human lineage to produce our chromosome number 2. There is no knowing at which point along our own lineage this chromosome fusion occurred, but if it happened *after* the split between the lines that became Cro-Magnon and Neanderthal then there would be a chromosome imbalance, with Neanderthals having forty-eight chromosomes and Cro-Magnons only forty-six. The offspring of a mating between a Cro-Magnon and a Neanderthal would have forty-seven chromosomes and, although it may have been completely healthy, it would find itself in the same difficulty as the mule when it came to producing sperm or eggs. No-one knows how many chromosomes the Neanderthals had, but I suspect one day we will be able to find out. I think the experiment could be done. Until then we won't know whether the complete absence of Neanderthal mitochondrial DNA in modern Europe is attributable to a fundamental biological or social incompatibility between our Cro-Magnon ancestors and the other human species with which they shared the continent.

The publication of our genetic conclusion about the extinction of the Neanderthals met with a tongue-in-cheek chorus of disbelief from the British tabloids. The *Daily Express* published a picture of a Neanderthal alongside a photograph of a characteristically sullen Liam Gallagher, the Oasis singer. How, it asked, could geneticists possibly claim that the Neanderthals were extinct when faced with such overwhelming evidence that they were alive and well in late twentieth-century Britain? Of course, they were predictably playing on the stereotype of Neanderthal as brutish and subnormal, for which there is no evidence at all. It was this kind of prejudice which dissuaded me from following up the several calls and letters I had from people who were sure that someone they knew (never themselves, of course) was definitely a Neanderthal. I still remember the letter from Larry Benson from Santa Barbara in California who wrote to tell me that a checkout clerk at his local supermarket had all the features of a Neanderthal. Apparently he was a really nice man, who (my correspondent assured me) would be only too pleased to provide a sample for DNA testing. I didn't take it up.

So the Neanderthals are extinct: completely replaced in Europe, and throughout their range, by the technologically and artistically superior new species *Homo sapiens*, represented in Europe by the Cro-Magnons. What happened in Europe, as far as we can tell from the genetics, also happened throughout the world, with *Homo sapiens* becoming first the dominant then the only human species, completely eliminating earlier forms. The Neanderthals, or *Homo neanderthalensis* as we are justified in calling them now that we are satisfied they constitute a separate species from our own, disappeared from Europe, and *Homo erectus* vanished from all of Asia. Whether *Homo sapiens* and *Homo erectus* ever overlapped in Asia is uncertain. In China there is a gap in the fossil record between 100,000 and 40,000 years ago. Perhaps *Homo erectus* had already died out before *Homo sapiens* arrived. There is no fossil evidence that *Homo erectus* ever reached Australia or the Americas, suggesting that *Homo sapiens* may have been the first humans to inhabit these two continents. In Africa, where *Homo sapiens* as a species first evolved, the equivalent replacement of other humans may have been sudden or gradual. Whatever the mechanism and whatever the reason, *Homo sapiens* has completely replaced other human species throughout the world. When the last Neanderthal died, twenty-eight thousand years ago, there was only one human species left to rule the planet. Ours.

There are no clear signs of interbreeding, no convincing remnants of earlier genes from these subdued species anywhere. But, as with the Europeans, so much remains untested. Who knows what the next sample will bring? Who can be sure that in the remote mountains of Bhutan, the lonely deserts of Arabia, the forests of central Africa or the crowded streets of Tokyo there is not a single person who holds the evidence of a different history embedded somewhere in his or her genes?

10

HUNTERS AND FARMERS

Although Cro-Magnon stone technology was a significant advance over the existing apparatus of the Neanderthals, life in the Old Stone Age was still based on hunting. Archaeologists divide the Stone Age into three phases, based on the style of stone tools used. It is not a hard and fast classification and is fuzzy at some of the boundaries, but it has endured as a useful way of referring to the main features of an archaeological site where the only evidence to go on is the artefacts that are found there. A trained archaeologist can tell at a glance whether he or she is dealing with an Old, Middle or New Stone Age site by the features of the stone tools and other artefacts found there and without needing to find any human bones to help.

The Old Stone Age, or *Palaeolithic* (from the Greek for *old* and *stone*) covers the time from the first appearance of stone tools about two million years ago up until the end of the last Ice Age about fifteen thousand years ago. There are huge differences between the crude hand axes that come from the beginning of this period and the delicately worked flint tools that are found at the end. To differentiate the various phases of this development, the Palaeolithic is divided into Lower, Middle and Upper phases. The Lower Palaeolithic roughly coincides with the time of *Homo erectus*, the Middle Palaeolithic corresponds approximately with the time of the Neanderthals, and the most recent, the Upper Palaeolithic, refers to the period beginning in Africa about a hundred thousand years ago when *Homo sapiens* finally arrived on the scene. In Europe, the Upper Palaeolithic doesn't begin until the first *Homo sapiens*, the Cro-Magnons, appear with their advanced stone technology between forty and fifty thousand years ago.

After the end of the last Ice Age, the Middle Stone Age, or *Mesolithic*, takes us up to the beginnings of agriculture. The boundary between the Upper Palaeolithic and the Mesolithic is very indistinct. There is an increase in the sophistication of worked stone tools and characteristic styles of implements made from bone and antler. Many more sites are found around coasts. However, there is no entirely new stone technology on the scale of that which divides the Middle and Upper Palaeolithic. At the other end of the Mesolithic, though, the transition is dramatic. The New Stone Age or *Neolithic* is the age of farming, and it is associated with a whole new set of tools – sickles for cutting stands of wheat; stones for grinding the grain – and, almost always, the first evidence of pottery.

The Cro-Magnons of the European Upper Palaeolithic lived in small nomadic bands following the animals they hunted, shifting camp with the seasons.

Although a vanishingly few people around the world still make a living like this, for most of us (certainly for most of you reading this book) the fundamental basis of life has changed dramatically. This is due to the one technical revolution which eclipses any refinements to the shape and form of stone tools in its importance in creating the modern world. That revolution is agriculture. Within the space of only ten thousand years, human life has changed beyond all recognition, and all of these changes can be traced to our gaining control of food production.

By ten thousand years ago, our hunter–gatherer ancestors had reached all but the most inaccessible parts of the world. North and South America had been reached from Siberia. Australia and New Guinea were settled after significant sea crossings, and all habitable parts of continental Africa and Europe were occupied. Only the Polynesian islands, Madagascar, Iceland and Greenland had yet to experience the hand of humans. Bands of ten to fifty people moved about the landscape, surviving on what meat could be hunted or scavenged and gathering the wild harvest of seasonal fruits, nuts and roots. Then, independently and at different times in at least nine different parts of the world, the domestication of wild crops and animals began in earnest. It started first in the Near East about ten thousand years ago and, within a few thousand years, new centres of agriculture were appearing both here and in what are now India, China, west Africa and Ethiopia, New Guinea, Central America and the eastern United States. This was not a sudden process but, once it had begun, it had an inexorable and irreversible influence on the trajectory of our species.

There has never been a completely satisfactory explanation of why agriculture began when it did and how it sprang up in different parts of the world during a period when there was no realistic possibility of contact between one group and another. This was a time when the climate was improving, though fitfully, after the extremes of the last Ice Age. It was becoming warmer and wetter. The movement of game animals became less predictable as rainfall patterns changed. Even so, none of these things in themselves explain the radical departure from life as a hunter to that of a farmer. Why hadn't it happened before? There were several warm interludes between Ice Ages during the course of human evolution where the climate would have favoured such experimentation. What must have been lacking was the mind to experiment.

Whatever the reasons behind the invention of agriculture, there is no doubting its effect. First of all, numbers of humans began to increase. Very roughly, and with wide variations depending on the terrain, one hunter–gatherer needs the resources of 10 square kilometres of land to survive. If that area is used to grow crops or rear animals, its productivity can be increased by as much as fifty times. Gone is the necessity for seasonal movements to follow the game or wild food. Very gradually camps became permanent, then in time villages and towns grew up. Soon food production exceeded the human effort available to keep it going. It was no longer essential for everybody to work at it full-time; so some people could turn to other activities, becoming craftsmen, artists, mystics and various kinds of specialists.

But it was not all good news. The close proximity of domesticated animals and dense populations of humans in villages and towns led to the appearance of epidemics. Measles, tuberculosis and smallpox crossed the species barrier from cattle to humans; influenza, whooping cough and malaria spread from pigs, ducks and chickens. The same process continues today with AIDS and BSE/CJD. Resistance to these diseases slowly improved in exposed populations, and here they became gradually less serious. But when the pathogens encountered a population which had not previously been exposed to them, they exploded with all their initial fury. This pattern would be repeated throughout human history. The European settlement of North America which followed on from the voyage of Christopher Columbus in 1492 was made easier by the accidental (or sometimes deliberate) infection of native Americans by epidemic diseases, like smallpox, which killed millions.

The first nucleus of domestication that we know about appeared about eleven thousand years ago in the Near East, in what is known as the Fertile Crescent. This region takes in parts of modern-day Syria, Iraq, Turkey and Iran, and is drained by the headwaters of the Tigris and Euphrates rivers. Here or hereabouts the hunters first began to gather in and eat the seeds of wild grasses. They still depended on the migrating herds of antelope that criss-crossed the grasslands on their seasonal migrations, but the seeds were plentiful and easy to collect. This was not agriculture, just another aspect of gathering in the wild harvest. Inevitably some seeds were spilled, then germinated and grew up the following year. It was a small step from noticing this accidental reproduction to deliberate planting near to the camps, which had by then already become more or less permanent in that part of the world thanks to the local abundance of wild food. Over time, the plants which produced the heavier grains were deliberately selected, and the natural genetic variants that produced them increased in the gene pool. True domestication had begun.

The same process was repeated in other parts of the world at later times and with different crops: rice in China, sugar-cane and taro in New Guinea, teosinte (the wild ancestor of maize) in Central America, squash and sunflower in the eastern United States, beans in India, millet in Ethiopia and sorghum in west Africa. Not only wild plants, but wild animals too were recruited into a life of domestication. Sheep and goats in the Near East along with cattle, later separately domesticated in India and Africa; pigs in China, horses and yaks in central Asia, and llamas in the Andes of South America were all tamed into a life of service. Even though most species resisted the process – for example, no deer have, even now, been truly domesticated – the enslavement of wild animals and plants for food production was the catalyst that enabled *Homo sapiens* to overrun and dominate the earth.

But how was this accomplished? Was there a

replacement of the hunter–gatherers by the farmers, just as the Neanderthals had been pushed aside by the technologically advanced Cro-Magnons? Or was it instead the *idea* of agriculture, rather than the farmers themselves, which spread from the Near East into Europe? This seemed like another case of rival theories that could be solved by genetics – so we set out to do just that.

By the summer of 1994, by which time I had secured the three-year research grant I needed to carry on, I had collected together several hundred DNA sequences from all over Europe, in addition to the samples we had acquired on our Welsh trip two years previously. Most of them had been collected by the research team, or through friends, as the opportunity arose. One friend of mine was engaged to a girl from the Basque country in Spain, so he surprised his future in-laws by arriving with a box of lancets and setting about pricking the fingers of friends and family alike. A German medical student who was spending the summer in my lab on another project went para-gliding in Bavaria and tucked the sampling kit into his rucksack. Other DNA samples came from like-minded colleagues in Germany and Denmark who sent small packages containing hair stuck to bits of sellotape. Hair roots are a good source of DNA, but they are fiddly to work with and a lot of people, especially blondes, have hair that breaks before the root comes out. And pulling out hair hurts.

Another year on, and by the early summer of 1995 a few papers were beginning to appear in the scientific literature on mitochondrial DNA, from places as far apart as Spain, Switzerland and Saudi Arabia. It is always a precondition of publication in scientific journals that you deposit the raw data, in this case the mitochondrial sequences, in a freely accessible database; so with the help of these reports we were able to build up our numbers of samples further. The papers themselves were not encouraging. Their statistical treatment of the data was largely limited by the computer programs available at the time to comparisons of one population average against another, and drawing those wretched population trees. Given this treatment, the populations looked very much like one another, and the authors inevitably concluded with pessimistic forecasts about the value of doing mitochondrial DNA work in Europe at all. Compared to the genetic dramas being revealed in Africa, where there were much bigger differences between the DNA sequences from different regions, Europe was starting to get a reputation for being dull and uninteresting. I didn't think that at all. There was masses of variation. We rarely found two sequences the same. What did it matter if Africa was 'more exciting'? We wanted to know about Europe, and I was sure we could.

When we had gathered all the European data together, we started by trying to fit the sequences into a scheme which would show their evolutionary relationship to one another. This had worked very well in Polynesia, where we saw the two very distinct clusters and went on to discover their different geographical origins. We soon found out that it was going to be much more difficult than that in Europe. When we plugged the data into a computer program that was designed to draw evolutionary trees from molecular sequences, the results were a nightmare. After thinking for a very long time, the computer produced thousands of apparently equally viable alternatives. It couldn't decide on the true tree. It looked hopeless. This was a very low point. Without a proper evolutionary scheme that connected the European sequences, we were going to be forced to publish our results, the results of three years' hard work and a lot of money, with only bland and, to me, pretty meaningless population comparisons that might conclude, say, that the Dutch were genetically more like the Germans than they were like the Spanish. Wow.

Before going down that miserable route – and we had to publish something soon to have any hope of securing yet more funding – we went back to the raw data. Instead of feeding it into the computer, we started drawing diagrams on bits of paper. Even then we couldn't make any sense of the results. For instance, we would have four obviously related sequences but couldn't connect them up in an unambiguous evolutionary scheme. Figure 4a shows an example. Sequence A was our reference sequence, sequence B had one mutation at position 189 and sequence C had one mutation at position 311. That's easy enough. Sequence A came first, then a mutation at 189 led to sequence B. Similarly, a mutation at 311 turned sequence A into sequence C. No real problem there. No ambiguity. But what to do with a sequence like D, with mutations at 189 *and* 311? D could have come from B with a

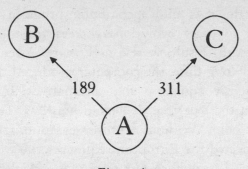

Figure 4a

mutation at 311, or from C with a mutation at 189 (see Figure 4b). Either way it was obvious that the mutations, on which everything depended, were happening more than once. They were recurring at the same position. No wonder the computer was getting confused. Unable to resolve the ambiguity, it would draw out *both* trees. Another ambiguity somewhere else would force the program to draw out four trees. Another one and it had to produce eight trees, and so on. It was easy to see that it wouldn't take many recurrent mutations in such a large set of data for the computer to produce hundreds or even thousands of alternative trees. How were we going to get over this? It looked as if we were really stuck. For the next week I would think I'd solved it, get out a piece of paper and start drawing, then realize whatever idea I'd had wasn't going to work. Finally, I was sitting down in the coffee room one day doodling on napkins when the solution dawned on me. Don't even try to come out with the perfect tree. Leave the ambiguities in there. Instead of

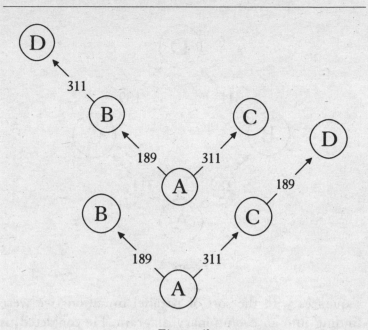

Figure 4b

trying to decide between them, just draw it as a square (Figure 4c). Freely admitting that I didn't know which route led to D, I could leave it at that. Once I had unhooked myself from this dilemma, the rest was easy. I could relax. I was no longer seeking the perfect tree from thousands of alternatives. There was just one diagram, not a tree but a network, which certainly included some ambiguities but whose overall shape and structure was full of information.

Unknown to our team in Oxford, a German mathematician, Hans-Jürgen Bandelt, had been working on the theoretical treatment for just such a scenario. He was looking for the best way of incorporating DNA

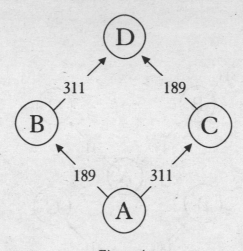

Figure 4c

sequences with the sort of parallel mutations we were finding into an evolutionary diagram. He contacted us because he needed some real data to chew over, and we at once realized that we were both thinking along the same lines and solving the problem in the same way, drawing networks and not trees. The big difference was that Hans-Jürgen was able to apply a proper mathematical rigour to the process of constructing the networks, an advantage which was crucial to their acceptance as a respectable alternative to the traditional trees.

With this important obstacle overcome, we could now concentrate on the picture that was slowly emerging from the European sequences. Whereas in Polynesia we saw two clearly differentiated clusters, in Europe the networks were sorting themselves out into several related clusters, groups of mitochondrial sequences that looked as though they belonged together. These were not so obviously distinct or so far apart as their Polynesian equivalents, in the sense that each cluster had fewer mutations separating it from the others. We had to look hard to make out the boundaries, and Martin Richards and I spent many hours trying to decide how they best fitted together. Were there five or six or seven clusters? It was hard to decide. At first we settled on six. We found out later that we had missed a clue which would have divided the biggest of the six clusters into two smaller ones to give us the seven clusters that we now know form the framework for the whole of Europe.

What mattered to us at the time was not so much precisely how many clusters there were, but that there were clusters at all. This was not the homogeneous and unstructured picture presented by the scientific articles which had been published by the summer of 1995, leading their authors to despair that anything useful could be found out about Europe from mitochondrial DNA. The clusters might have been hard to see, in fact impossible to distinguish without the clarifying summary of the network system, but there was no mistaking their presence. Now that we had our seven defined clusters we knew what we were dealing with, and could start looking at where they were found, and how old they were. Because we had a figure for the mutation rate of the mitochondrial control region we could combine this with the number of mutations we saw in each of the seven clusters to give us an idea of how long it had taken each cluster to evolve to its current stage of complexity. This had worked beautifully in Polynesia, where the two clusters we found had accumulated relatively few mutations within them for the simple reason that humans had only been in Polynesia for three to four thousand years at most. When we worked out the genetic dates for the two Polynesian clusters in the different island groups by factoring in the mutation rate, they corresponded very well to the settlement dates derived from the archaeology. The earliest islands to be settled, Samoa and Tonga in western Polynesia, had the most accumulated mutations within the clusters and a calculated genetic age of three thousand years, very similar to the archaeological age. Further east, the Cook Islands had fewer accumulated mutations and a younger date. Aotearoa (New Zealand), the last Polynesian island to be settled, had very few mutations within the clusters and the youngest date of all.

When we applied exactly the same procedure to the clusters in Europe we got a surprise. We had been expecting relatively young dates, though not as young as in Polynesia, because of the overwhelming influence of the agricultural migrations from the Near East in the last ten thousand years that were so prominent a feature in the textbooks. But six out of the seven clusters had genetic ages much older than ten thousand years. According to the version of Europe's genetic history that we had all been brought up on, a population explosion in the Near East due to agriculture was followed by the slow but unstoppable advance of these same people into Europe, overwhelming the sparse population of hunter–gatherers. Surely, if this were true, the genetic dates for the mitochondrial clusters, or most of them at least, would have to be ten thousand years or less. But only one of the seven clusters fitted this description. The other six were much older. We rechecked our sequences. Had we scored too many mutations? No. We rechecked our calculations. They were fine. This was certainly a puzzle; but still we didn't question the established dogma – until we looked at the Basques.

For reasons discussed in an earlier chapter, the Basques have long been considered the last survivors of the original hunter–gatherer population of Europe. Speaking a fundamentally different language and living in a part of Europe that was the last to embrace agriculture, the Basques have all the hallmarks of a unique population and they are proud of their distinctiveness. If the rest of Europe traced their ancestry back to the Near Eastern farmers, then surely the Basques, the last survivors of the age of the hunter–gatherers, should have a very different spectrum of mitochondrial sequences. We could expect to find clusters which we saw nowhere else; and we would expect not to find clusters that are common elsewhere. But when we pulled out the sequences from our Basque friends, they were anything but peculiar. They were just like all the other Europeans – with one noticeable exception: while they had representatives of all six of the old clusters, they had none at all of the seventh cluster with the much younger date. We got hold of some more Basque samples. The answer was the same.

Rather than having very unusual sequences, the Basques were as European as any other Europeans. This could not be fitted into the scenario in which hunters were swept aside by an incoming tide of Neolithic farmers. If the Basques were the descendants of the original Palaeolithic hunter–gatherers, then so were most of the rest of us.

But what about the cluster that was absent from the Basques – the cluster that was distinguished from the rest by having a much younger date compatible with the Neolithic? When we plotted the places where we found this cluster on a map of Europe, we found a remarkable pattern. The six old clusters were to be found all over the continent, though some were commoner in one place than in others. The young cluster, on the other hand, had a very distinctive distribution. It split into two branches, each with a slightly different set of mutations. One branch headed up from the Balkans across the Hungarian plain and along the river valleys of central Europe to the Baltic Sea. The other was confined to the Mediterranean coast as far as Spain, then could be traced around the coast of Portugal and up the Atlantic coast to western Britain. These two genetic routes were exactly the same as had been followed by the very first farmers, according to the archaeology. Early farming sites in Europe are instantly recognizable by the type of pottery they contain, just as Lapita ceramics identify the early Polynesian sites in the Pacific. The push through central Europe from the Balkans, which began about seven and a half thousand years ago, is recorded by the presence at these early sites of a distinctive decorative style called Linear pottery, in which the vessels are incised with abstract geometric designs cut into the clay. The Linear pottery sites map out a slice of central Europe where, even today, one branch of the young cluster is still concentrated. In the central and western Mediterranean, early farming sites are identified by another style of pottery, called Impressed ware because the clay is marked with the impressions of objects, often shells, which have been pressed into the clay before firing. Once again, the concordant distribution of Impressed ware sites and the other branch of the young cluster stood out. This didn't look like a coincidence. The two branches of the young mitochondrial cluster seemed to be tracing the footsteps of the very first farmers as they made their way into Europe.

There was one further piece of evidence we needed before we could be confident enough to announce our radical revision of European prehistory to the world. If the young cluster really were the faint echo of the early farmers, then it should be much commoner in the Near East than it is in Europe. At that time, the only sequences we had available from this region were from the Bedouin of Saudi Arabia. While only about 15–20 per cent of Europeans belonged to the young cluster – depending on which population was being studied – fully half of the Bedouin were in it.

We now had the evidence that most modern Europeans traced their ancestry back, far beyond the Neolithic, to the hunter–gatherers of the Palaeolithic, including the first Cro-Magnons that had replaced the

Neanderthals. Certainly there had been new arrivals from the Middle East in the Neolithic; the correspondence between the geographical pattern of the young cluster and the archaeologically defined routes followed by the early farmers was good evidence of that. But it was not an overwhelming replacement. The young cluster makes up only 20 per cent of modern Europeans at the very most. We were ready to go public.

11

WE ARE NOT AMUSED

Professor Luigi Luca Cavalli-Sforza is a man whose eminence is matched only by his elegance. Erect of posture, even in his late seventies, his silver hair always immaculately groomed, he is equally at home in the busy conference rooms of the academic circuit by day and the exclusive restaurants that welcome the most distinguished delegates by night. His contributions and influence in the field cannot be over-estimated. Scientists who once studied under him, either in Italy or later at Stanford University in California, today hold many of the important academic positions in the discipline of human population genetics. It was Luca who first formulated the theory which had come to dominate European prehistory over the preceding quarter-century. According to this theory, or at least the version believed by archaeologists, farmers from the Near East had overwhelmed the descendants of the Cro-Magnons, who themselves had replaced the Neanderthals. This was a large-scale replacement which meant that most Europeans traced their ancestry back not to hunter–gatherers but to farmers.

Having collected together the records of thousands of blood and other genetic tests from all over Europe, Luca had amalgamated the results into a gradient of gene frequencies that summarized this mountain of data. These gradients were organized into simple vectors, called principal components, which were projected as lines on a map. The most striking, the first principal component, led diagonally across Europe from Anatolia in Turkey to Britain and Scandinavia in the north-west. To Luca and his colleagues, this was the signature of a massive influx of people into Europe from the Near East. The fit between the south-east/north-west axis of this genetic slope and the routes followed by the early farmers according to the archaeology available at the time was convincing. The farmers had overrun Europe.

The influence of Cavalli-Sforza's conclusion spread far beyond the narrow bounds of human genetics, through archaeology and related disciplines. Although there were some archaeologists who did not share his conclusion and saw in the record signs of only minimal population movement, they had a hard time making their views heard. Like any academic discipline archaeology has its fashions, and the fashion in Europe was for a large-scale settlement by incoming farmers. It had not been so when Cavalli-Sforza and his colleague, the American archaeologist Albert Ammerman, first put forward their ideas in the 1970s. At that time the contemporary taste was for entirely indigenous development; for the gradual adoption of agricultural methods and practice by the mesolithic hunter–

gatherers of Europe without a large-scale movement of people. The original argument put forward by Ammerman and Cavalli-Sforza was for at least some movement, some migration from the Near East. Launched in a hostile intellectual atmosphere, this process was described in a term which sounded un-challenging. It was called 'demic diffusion'. Demic means 'to do with people', and diffusion is a gentle phrase implying the gradual inching outward of the farmers from their stronghold in the Near East. How-ever, demic diffusion was not just a descriptive idea; it had a strong mathematical basis. It took as its foundation a mathematical model developed by Arthur Mourant's mentor, the great statistical geneticist R. A. Fisher, who produced equations to describe the spread of anything – animals, people, genes, ideas – outwards from a growing centre. This mathematical model was given the dramatic title the 'wave of advance'.

Over the past twenty-five years, the 'wave of advance', the name of the mathematical model, has gradually taken over from 'demic diffusion' as the description of the spread of farming. I don't entirely understand the reasons for this. It may be that as the model became more widely accepted there was no need to present it in a tone conciliatory to the intellectual atmosphere into which it was introduced, resistant to any theories which suggested large-scale movements of people; or it may just have been that archaeologists were beguiled by the power of the phrase 'wave of advance'. In any event, somehow the dramatic had taken over from the gentle. The idea of a gradual influence of incoming agriculturalists had been replaced in the collective psyche by the image of an unstoppable tidal wave of land-grabbing farmers that swept away everyone and everything in its path. The notion that the farmers overwhelmed the original inhabitants became the prevailing lore among archaeologists.

Not only had this tsunami of people brought agriculture to Europe, it was also responsible, according to the distinguished Cambridge archaeologist Colin Renfrew, for the introduction and dissemination of the language family to which most European languages belong. Although it is not readily apparent to any but professional linguists, there is no doubt that, with only a few exceptions, the languages spoken in Europe today all stem from a common root. They belong to a family of languages called Indo-European. The way in which sentences are constructed and many of the words they share betray a relationship among them that may not be obvious to most of us as we struggle with our phrase books. It takes a linguist to connect English and Portuguese, Greek and Gaelic. The exceptions are the Basques' Euskara, Finnish, Estonian, Lapp and Hungarian. While Euskara is unique among living European languages and cannot be reliably linked to any other (though some linguists see a connection with languages of the Caucasus mountains), the other four are members of the Uralic language family which has its origins further east.

The Indo-element in Indo-European is there because there is a strong connection, again visible only to linguists, between the European languages and Sanskrit. This link was discovered by William Jones in 1786 while he was working as a judge in India for the British Raj. It was an amazing piece of amateur scholarship; indeed, Jones invented the concept of language families that is still a feature of comparative linguistics today. The essential idea of a language family is that all the different languages within it have evolved from a common root, almost certainly a language that is by now extinct. This raises the question of where the original Indo-European language was spoken and, importantly, how it spread out from there. Renfrew deduced that the original Indo-European language was spoken in Anatolia in central Turkey, and was then spread to Europe by the first farmers. A massive replacement of the hunter–gatherers by the agricultural 'wave of advance', as demic diffusion had surreptitiously become, was just what was needed to spread the language from its base in Anatolia.

There was now a powerful confederation of genetics, archaeology and linguistics in support of the argument that the mesolithic hunter–gatherers of Europe had been overwhelmed by the neolithic farmers. So, by the time we produced our startling results, the received wisdom was that most native Europeans today were descended not from the people who had endured the rigours of the last Ice Age but from the farmers who had walked in only ten thousand years ago with a bag of seeds and a few animals. But it just didn't fit with the ages of our DNA clusters. We were sure that the strongest signals from the mitochondrial DNA in today's Europeans were from much further back in the past than ten thousand years. We saw these signals as the genetic echo of the hunter–gatherers. These were not the faint whispers of a defeated and sidelined people but a resonant and loud declaration from our hunter–gatherer ancestors: 'We are still here.'

I decided to present our work at the Second Euro-conference on Population History, held in Barcelona in November 1995. I knew very well that the main proponents of the 'wave of advance' theory would be there, so at least what I had to say would be noticed. I was given a twenty-minute slot. The conference room was vast, with four hundred delegates and room for many more. I was introduced by the convenor, Sir Walter Bodmer, Fellow of the Royal Society, a long-time associate of Luca Cavalli-Sforza and co-author with him of two influential textbooks on genetics. Walter is not widely known for his conciliatory remarks, but I did think 'And the next speaker is Bryan Sykes who is talking about mitochondria. I don't believe in mitochondria' was a less than gracious intro-duction. I began to lay out the basis for our revision of European prehistory.

Walter and Luca were both sitting below the podium, side by side in the front row. It is surprising how much you can take in when addressing even a large audience such as this. As I went from one point to another I could see that Walter was getting agitated. He began to mutter to himself, then to Luca; at first inaudibly, then louder and louder. 'Rubbish,' 'Nonsense,' I thought I heard him say. He began to fidget, to half raise himself in his seat then sit back down, as one slide followed another in my presentation. As I came to the concluding slide, I could almost see the steam coming out of his ears.

No sooner had I finished talking than Walter and Luca were on their feet, throwing questions at me. I have known Walter for ages and seen him in action many times. I have watched him crush young researchers by his aggressive questioning, and I was determined the same would not happen to me. There is only one effective remedy with Walter, and that is to argue back. I had been expecting fireworks, and as I stood there under this barrage, I began to see it all as a piece of theatre – like a cross-examination in the High Court or a fierce exchange at the Despatch Box in the House of Commons. I began to enjoy myself.

At one point Walter insisted that they (he and Luca) had never said that the farmers had overwhelmed Europe and replaced the hunter–gatherers. I had brought along a copy of their jointly written textbook *Genetics, Evolution and Man* against just such an assertion. In response, I opened it at a page I had already marked with a yellow sticker and read out: 'If the population of Europe is largely composed of farmers who gradually immigrated from the Near East, the genes of the original Near Easterners were probably diluted out progressively with local genes as the farmers advanced westward. However, the density of hunter–gatherers was probably small and the dilution [of Near Eastern genes, that is] would thus be relatively modest.' There it was in black and white, in their own words.

This was massive replacement in all but name. Walter puffed one last time and sat down. The chairman closed the session. I had survived the first charge: but the fuse had been lit on a fierce debate that would not be resolved for another five years.

In science these days, international conferences like the one in Barcelona are useful for announcing new findings and getting an initial reaction. But work presented at a conference has no real validity until it is published in a scientific journal. Publication involves close scrutiny of the data, the methods and the conclusions by expert reviewers working unpaid and under an obligation to declare any conflict of interests. Though a conference presentation has to be truthful, it is only during the review process prior to publication that the assumptions, results and interpretations are thoroughly checked. Considering the fierce reaction that our radical revision of European prehistory had provoked in Barcelona, it came as no surprise to us when we submitted our manuscript to the *American Journal of Human Genetics*, the leading international journal in the field, that the reviewers were even more demanding than usual. They insisted that the evolutionary network method, which we had published in 1995 as an intensely mathematical and opaque article, be explained once more in an appendix. They asked for additional tables of, to my mind, old-fashioned population comparisons. But finally, they published it. 'Palaeolithic and Neolithic lineages in the European mitochondrial gene pool' appeared in the July 1996 issue. It was now in print. We had set out our stall; now we waited for the reaction.

For a while, nothing happened. Then we started hearing from friends that the work was being discussed as at best irrelevant or at worst just plain wrong. Surprisingly, the main target of the whispering campaign was not us but mitochondrial DNA itself, which had distinguished itself so well in solving the puzzle of the Polynesians. Suddenly it was portrayed as being unreliable, too unstable, with too many parallel mutations in the section that we had chosen to use. The mutation rate estimates were attacked as being wildly out. This meant that the dates for the clusters were much younger than we thought and thus perfectly compatible with the 'wave of advance' model of an essentially farming-derived gene pool. Lastly, mitochondrial DNA was accused of being just one marker, just a single witness to events whose account of prehistory could not be substantiated.

When a controversial paper is published it is not unusual for the scientific journal in which it appears to receive and publish a criticism from others in the field. This takes the form of a 'Letter to the Editor'. The authors of the original paper are given the opportunity to respond, and if they do, both letters appear next to each other in the same issue of the journal. It was no surprise to learn that Cavalli-Sforza had composed such a criticism of our paper and that it had been accepted by the *American Journal of Human Genetics*. The editor sent us a copy of Luca's letter with an invitation to reply to it.

The letter was a withering attack on mitochondria in general and on our interpretation of the control region sequence data in particular. It did, however, contain one very interesting statement that we had been waiting to hear. Although the overwhelming influence of the neolithic farmers on the make-up of the European gene pool was the main feature of Luca's 'demic diffusion/wave of advance' model, no figures had ever been put on their overall genetic contribution. While we had estimated that roughly 20 per cent of modern Europeans traced their mitochondrial ancestors back to these early agriculturalists, there was no comparable figure from Luca's work that we could use as a contrast. The assumption which most people had made was that the farmers had 'overwhelmed' the hunters. That was certainly how a generation of archaeologists had interpreted the 'wave of advance' model. But the scale of the immigration had never been quantified. There was probably no need. The model had gained its own momentum and everybody knew what it meant, or thought they did. But now, for the first time, Luca put a figure on the proportion of modern European genes contributed by farmers from the Near East. It was, according to the letter, roughly equal to the proportion of the genetic variation that contributed to the first principal component which tracked the cline of genes across Europe from the south-east to the north-west. And this was 26 per cent. No mathematical proof whatsoever accompanied the statement, but we weren't going to complain about that. It was close enough to our estimate of about 20 per cent, derived from mitochondrial DNA analysis, that it looked as if there was little left to argue about.

Even though this was an important new announcement from Luca, we certainly needed to reply to his letter and the criticism of mitochondrial DNA that it contained. He had every right to be critical. It is perfectly reasonable to demand absolute clarification from anybody who is challenging a long-held view. Extraordinary claims, such as ours, demand extraordinary proof. Even so, we all felt under a lot of pressure. We were the new boys on the block up against the might of the Establishment. Nevertheless, I never doubted for a second that we were right. There was nothing for it but to answer the criticisms one by one.

We were confident that the first objection – that our chosen section of mitochondrial DNA, the control region, was so riddled with parallel mutations as to be completely unreliable – could be rebuffed. There are plenty of other base changes that can be used as molecular markers around the mitochondrial DNA circle. If we drew a new evolutionary tree using these other markers instead of the control region sequences, then one of two things would happen: either the clusters would match our own groupings or they would not. If they did match, then the control region must be reliable. If they didn't match, then it wasn't, and we might as well give up.

For this test we teamed up with Antonio Torroni, an Italian geneticist from Rome who had spent many years developing an intricate technical system for these other markers. He supplied us with samples he had already tested for us to sequence through the control region, and we in turn took our own sequenced samples to

Rome to run through his system. The results couldn't have been more encouraging. There was an almost exact fit between the clusters identified by Antonio's markers and our own. The one or two minor incompatibilities were quickly resolved; those apart, the match was perfect – so much so, in fact, that we abandoned our own numerical classification for the clusters and adopted Antonio's, based on letters of the alphabet. Now we had proof that the control region was not after all a fickle piece of DNA that could mislead and deceive but, once you got to know it, a faithful and reliable companion.

The mutation rate criticism was harder to address. It was certainly true that if we were using a gross under-estimate of the mutation rate then our cluster dates would be seriously adrift. If our estimates were out by a factor of ten, as some people suggested, then the ages of our clusters would fall from the Palaeolithic into the Neolithic and we could kiss our theory good-bye.

There are basically two ways of estimating a mutation rate. Either you can try to measure it by direct observations from one generation to the next, or you can see how many mutations have accumulated in two different groups – which could be tribes, or populations, or species – that have been separated for a known length of time. The very first estimate of the mutation rate, the speed of the molecular clock, was made by comparing the differences between humans and their closest relative, chimpanzees, and combining this with the time since they last shared a common ancestor, estimated at between four and six million years ago. Of course, precisely when that separation between the ancestors of humans and chimpanzees took place is not known, especially since there are no chimp fossils to help out. The other route that has been used is to estimate the mutation rate changes which have accumulated in native Americans, who first arrived on the continent about twelve thousand years ago. The remarkable thing is that both methods agree so well with each other and come out with a figure of around one mutation in twenty thousand years down a single maternal lineage. When tracking back to a common ancestor between two modern people, as I did when estimating the date of the common ancestor between myself and the Tsar, there are *two* lineages, each with a chance to mutate, going forward from our common ancestor to each of us. Only one mutation separates my control region sequence from the Tsar's, but that mutation could have happened anywhere along the two maternal lineages leading from our common ancestor. At a rate of one mutation every twenty thousand years along a single lineage, that fixes the combined length of these two lineages to twenty thousand years. Since the Tsar and I are more or less contemporaries, the length of each lineage back to the common ancestor is therefore halved to ten thousand years. Our work in Polynesia had also shown an excellent agreement between the genetic and archae-ological dates for settlement using this mutation rate. If the rate was wrong by a factor of ten in Europe, then it had to be wrong everywhere else. It would mean that chimps and humans diverged only 400,000–600,000 years ago, America was first settled only 1,200 years ago and Polynesia only 300 years ago – in fact after the Europeans got there. This was so obviously crazy that the rates we were using couldn't be that far out.

Measuring mutation rates directly is a hard business. It means picking up a change between a mother and her child. We estimated that we would need to test a thousand pairs of parents and children to pick up a single new mutation. That was out of the question. Fortunately, the mutation process in mitochondria is a gradual one and, as it turned out, not too difficult to observe by a different route. Mutations happen in individual DNA molecules in individual mitochondria. However, in most people the DNA sequence of all the mitochondria in all the body cells is exactly the same. These two truths pose a paradox. A new mutation can only take place in one DNA molecule in one mitochondrion in one cell; so how does it manage to take over the whole body?

In order to be passed on to a new generation, a mutation has to occur in a female germline cell, one of the cells that divide to become eggs. Mutations also happen in other body cells – in skin, bone, blood, and so on – but, as these do not get passed on to the next generation, they play no part in the patterns of evolution. What seems to be happening is that each time a female germline cell divides it takes only a few mitochondria with it. If the mitochondrion with the new DNA mutation is one of the few to slip through this bottleneck then it can make up a much bigger proportion of the mitochondrial DNA in the new cells. When these cells divide there is a chance that the new mutation will be further enriched, and so on.

There are only twenty-four cell divisions in the female germline between one generation and the next. These are twenty-four opportunities for enrichment of a new mutation; only rarely is this enough for a complete takeover in a single generation. The individual who grows from the fertilized egg will have a mixture of two mitochondrial sequences: the old one, which is the same as her mother's, and the new one, which began as a new mutation somewhere in her mother's germline cells.

We looked very hard at our sequencing results over the past few years, searching for the signs of mixed mitochondria within the same person. We found that about 1.5 per cent of people do indeed have a mixture of two different mitochondrial DNAs. We then tracked these mixtures through families and found that it took an average of six generations for a new mutation to establish itself and take over completely. Remember the unusual case of the Tsar, who had a mixture of two different mitochondria in his bone cells? It looks as if he was in the transitional state where a new mutation was struggling to get established; eventually it did, as we can see in the cells of his modern-day relatives like Count Trubetskoy. As far as we could tell from our experiments, there was no inevitability in this process; some new mutations appeared to be doing well for one or two generations, then slipped back into obscurity and disappeared. We were observing directly the appearance and spread of new mutations, and from these data we could make a separate estimate of the mutation rate, independent of the complications associated with the exact dating of past events like the evolutionary separation of humans and chimps. This independent estimate, though only approximate, matched the mutation rate we had been using. We had answered the second criticism. Mitochondrial DNA had survived with its reputation intact.

The points Luca had raised in his letter, and to which we had responded, were serious and valid questions to ask of a new technology, especially one that had rewritten the version of prehistory that had dominated thinking for so long. They needed to be addressed, and they were. What happened next threatened to discredit not only our studies in Europe but all the evolutionary work using mitochondrial DNA that had ever been done on humans. We had to deal with the spectre of recombination.

Briefly, what makes the chromosomes in the cell nucleus so difficult to use for tracing evolutionary histories is their habit of scrambling information at each generation. Until the germline cells are into their final division which produces the gametes (sperm or eggs), the chromosomes lead separate lives and don't have a great deal to do with one another. However, in that final cell division, the pairs of chromosomes which have been inherited from each parent sidle up to one another, like mating earthworms, and start to exchange bits of DNA. After this canoodling they pull apart and go off to different gametes. But now they are no longer the same chromosomes but DNA mosaics. They have undergone what is called *recombination*. This is the ultimate genetic reason for sex itself, the potential for creating through recombination new and better gene arrangements that can advance evolution.

Recombination has its advantages for scientists. It has greatly helped the mapping of genes for serious inherited diseases on to specific chromosomes, and has been instrumental in unravelling the sequence of the entire human genome. But as far as tracing DNA through the generations is concerned, recombination is a very big nuisance. One of the features of mitochondrial DNA that have made it such a successful instrument for probing into the deep human past is that the information it brings us is *not* scrambled by recombination. The only differences between my mitochondrial sequence and that of my direct maternal ancestors are the changes that have been introduced over the millennia by mutation. With recombination, there would be the prospect of having not just one line of mitochondrial ancestors but dozens of them. Everything that had been assumed about mitochondrial genetics would be in doubt.

So, when two papers claiming evidence for mitochondrial recombination appeared in the March 1999 issue of the prestigious *Proceedings of the Royal Society*, they sent shock waves around the world. Editorials in the leading popular science journals, *Science* in Washington and *Nature* in London, immediately publicized this fundamental challenge to the authority of mitochondrial DNA. If recombination really was occurring, as these papers were suggesting, then it meant that all the work published over the previous decade on mitochondrial DNA in human evolution was completely undermined.

The wide publicity accorded to these articles was due not only to the claims they advanced but also to the great distinction of the author of one of them: John Maynard Smith, the undisputed doyen of British evolutionary biologists, the author of textbooks and other influential works, and still an active presence in his eighties. Condemnation by such an eminent figure, with no obvious axe to grind, spelled obliteration for us and everybody else in the field – if the claims for recombination could be substantiated. The substance of Maynard Smith's largely theoretical argument was that there was too much variation in mitochondrial DNA to have arisen by mutation alone. It was not so much a proof of recombination as an elimination of other mechanisms that could account for what Maynard Smith saw as a higher than predicted number of mutations. The reasoning was reminiscent of Sherlock Holmes's advice to Dr Watson in *The Sign of Four*: 'When you have eliminated the impossible, whatever remains, *however improbable*, is the truth.' But what made Maynard Smith's argument so seductive was the announcement in an adjoining paper of actual evidence for recombination in mitochondria from the tiny and remote island of Nguna in the Pacific. And the leading author (one of six) of the second paper was Erika Hagelberg.

Erika, you will recall, had worked in my laboratory

on the first recovery of DNA from human bone back in the late 1980s. She had since made a name for herself in the field of ancient DNA and become involved in some celebrated forensic cases, most famously when she and her colleagues had recovered DNA from the remains of Joseph Mengele, the infamous Nazi doctor who carried out unspeakable human experiments on prisoners in the Auschwitz extermination camp. With these and other cases under her belt she had built up a reputation as an imaginative scientist. However, despite occasional attempts on both our parts to heal the rift that had grown up during the difficult final days Erika spent in my laboratory, she and I had endured an uneasy relationship ever since. This tension added an extra dimension to the drama that was about to unfold.

The essence of Erika's evidence for recombination was that a particular mitochondrial mutation, at position 76 in the control region, was cropping up in several different clusters on the small island of Nguna. Like the Maynard Smith paper that accompanied it, this wasn't direct evidence for mitochondrial recombination. However, mutations at position 76 were exceedingly rare elsewhere in the world, so to find it frequently *and* in different clusters on the same island did deserve a special explanation. It would mean either that the mutation had happened spontaneously several different times in different clusters, which was extremely unlikely, or that a new mutation at 76 in one cluster had somehow spread to the others. And the only way for that to happen was by recombination.

For mitochondrial recombination to occur, two things have to happen. First, there needs to be a way for two circular mitochondrial DNA molecules to snuggle up to each other and exchange DNA. That didn't seem too unlikely. There are about eight DNA molecules in each mitochondrion and they enjoy free access to each other. So it would not be hard for them to exchange DNA. More difficult to accept was that there had to be two very different mitochondrial genomes in the same cell. If all the mitochondria in the cell had exactly the same sequence, they could exchange DNA between themselves as much as they liked and it would not make any difference. All the mitochondria would still have the same DNA sequence. Only if there were two *different* mitochondria exchanging DNA would anything be noticed. So the Nguna observation demanded that there were, or had been in the past, people who had mixtures of mitochondria. One component of the mixture would have to be the DNA belonging to one cluster, let's call it A, and with a mutation at position 76 in the control region. The other would be mitochondrial DNA from a completely different cluster, which we can call B, without the mutation at position 76. These two mitochondria would then exchange segments of DNA so that a piece from A, which included the mutation at position 76, ended up on B.

There was only one way to get two mitochondria from completely different clusters in a cell: one of them had to be coming not from the egg but from the sperm. So, if this claim of recombination turned out to be true, it would be a lethal double blow. Not only would it be impossible to trace mitochondrial lineages back in time because of the scrambling implicit in recombination, but it would also follow that mitochondrial inheritance was not, after all, exclusively maternal. No longer would it be safe to assume that our mitochondrial DNA had come from an ancestral line of mothers. It could have come from fathers as well. Something had to be done. We held an emergency meeting.

Vincent Macaulay, who trained as a physicist and was a formidable mathematician, and had joined the team two years previously, went off to check and recheck the sequence data used in the Maynard Smith paper. Incredibly, a lot of them were wrong. Either they had been incorrectly copied from the public databases, or the raw sequences themselves which had been deposited in these databases had mistakes in them (actually a common enough occurrence). The cumulative effect of both sorts of error made it look as though there were more mutations in the mitochondria than there really were. After correcting these mistakes in the data and redoing the Maynard Smith calculations, it was obvious that the force of the theoretical argument for recombination was seriously diluted. We wrote at once to Maynard Smith, who gracefully accepted the error.

The claim for recombination advanced by Erika Hagelberg was a more serious proposition. Even though it fell short of an actual proof of recombination, which would require a definition of the segments that had been exchanged between the two different mitochondria, it was still a piece of evidence that was hard to explain by any other mechanism. As far as I could see, it could only be wrong if there had been a massive systematic error in the sequencing of the Nguna samples. This seemed very unlikely, given that Erika was an experienced scientist who would be familiar with the rule that extraordinary claims needed extraordinary proofs. Conventionally, these sequences would have been repeated and checked several times before making such a radical claim that she must have realized would have such profound implications.

Nguna itself is a tiny island lying off Espirito Santo in Vanuatu, west of Fiji, and Vanuatu was one of the island groups which we had included in our earlier work on Polynesia. We had been given a few samples and, checking back, I found that four of them came from Nguna itself. In those days we did not report mutations lower than position 93, because the systems we used at the time sometimes gave unreliable readings below that. So it was no surprise that our computer records showed no mutations at the crucial site of position 76. However, we still kept the old X-ray films on which the sequence was displayed as a series of bands. By some miracle I managed to locate the Nguna plate dated 2 June 1992, and the quality was perfect. I could easily read the sequence down to 76 and beyond. There was no sign of a change at 76 in any of the samples. I went at once to my colleague in the Institute who had supplied me with the original blood samples and explained what I had found. He had some more from Nguna, and we tested those for the change at 76. Not one of them had it. It seemed incredible that we couldn't find the 76 mutation in twenty samples from such a tiny island when Erika was reporting it in nearly half of hers from the same place.

The situation was serious enough to warrant contacting Erika, and I emailed her in Dunedin, New Zealand, where she had recently taken up a post at the University of Otago. Given our strained relationship, I was as diplomatic as possible and stuck to the point. I explained that we had found no sign of the crucial mutation at position 76 in samples from the same small island. Would she let me know the source of the relevant Nguna samples, and send me samples so that I could replicate her findings? She replied that she was sure of the sequences and would re-check the results as soon as she could, that the possibility of a sequencing mistake is always there but that she had been reassured by the sheer mass of data. Considering the gravity of the situation and the impact even the suspicion of mitochondrial recombination was having on the reputation of the field as a whole, I then made a second request for samples of the Nguna DNA. This is unusual but not unheard of. I mentioned earlier that whenever a scientific paper is published there is an implicit undertaking, where possible, to make the raw material available for verification. This principle is at the very foundation of scientific progress. Without independent verification, or at least the opportunity to do so, scientific results have no validity. In most cases an actual test is unnecessary because the findings are quickly overtaken by new results. But here we had a situation where an entire field had been threatened with extinction. The truth about the Nguna samples, whatever it was, had to come out. And quickly.

I am sad to report that my requests for samples to verify the Nguna sequences did not produce results. Nor did I know of other laboratories that had tried to contact Erika to replicate the results. In the meantime, the reputation of mitochondrial DNA as a reliable evolutionary tool was spiralling downwards. The undergraduates had heard all about it. In the 1999 biological anthropology exams at Oxford, the demise of mitochondria featured in many of the students' answers. At a packed meeting in the zoology department at which some new work from Maynard Smith was being presented by one of his colleagues, I found myself in the distinctly uncomfortable position, during questions at the end of the lecture, of having to defend the reputation of mitochondria in front of an audience of very distinguished and influential evolutionary biologists who seemed only too eager to write it off.

I was pretty sure by now that Erika's Nguna data were wrong. Still, it was no good my thinking that. It was not really much use publishing our own results from the same island, either. There would still be uncertainty, and the original paper would still stand. If it were wrong, then it had to be corrected in the scientific press by Erika herself. In the meantime, I had also contacted co-authors of the paper who cooperated as far as possible: but still no sign of the samples.

In September of 1999 there was to be a conference in Cambridge at which both Erika and I were down to speak. It was a conference about Europe, and I gave a paper early on about our European work. Erika had been invited to talk about the Pacific islands and, we all assumed, about mitochondrial recombination. Generally speaking, scientific conferences are intensely polite affairs. There is a brief introduction by the session chairman; the speaker comes to the front and presents the paper, usually illustrated by a few slides or overheads; there is polite applause, a few questions from the audience, perhaps a bit more applause; the chairman introduces the next speaker. On this occasion, by the time it came for Erika to speak, there was a tangible atmosphere of anticipation, the expectation of a show-down in the air. The audience was completely silent, not wanting to miss a single word.

Erika began by saying that she was not going to talk about recombination. A murmur of surprise spread round the audience. Why had she come halfway round the world to a meeting on the genetic history of Europe if not to talk about mitochondrial recombination? As she went through her text on other aspects of her work in the Pacific, I knew I had to ask her about her Nguna work during questions, even if it had not featured in the presentation itself. It was the only way to get the matter cleared up. Was she sticking to her story or not? As Erika finished speaking, I raised my hand and the chairman called me to put my question. I was very nervous indeed, and could feel my heart pounding. But the issue was so important that I pressed on, in as unemotional a tone as I could manage.

'Erika,' I began, 'although you did not refer to this specifically in your talk, there has been, as you know, considerable interest in your claim of finding examples of mitochondrial recombination on the island of Nguna. As you also know, my laboratory did not find evidence for recombination in samples from the same small island. There has been a suggestion in the scientific press [which there had, and not by me] that there may be a systematic error in the DNA sequences which appeared in the article. How do you respond to this suggestion?'

She answered instantly that she had checked the sequences and stood by them.

I had to keep going. 'In that case, Erika,' I replied, 'why have you refused my requests for samples of the original DNA so that the sequences could be independently verified?'

The entire conference hall froze into complete silence.

'I did not refuse,' she answered.

'But you did not reply to my request, which amounts to the same thing,' I argued.

This was turning into a Grade One row. Erika accused me of having not scientific but personal motives for pursuing the matter. Fortunately, before I could answer this charge, someone else asked a related question about the recombination data and got what seemed to me to be an equally unconvincing reply. And yet, though by now many in the audience must have had their doubts about her original paper, at the end of the meeting it was still standing. There was no retraction. Not yet.

After that conference, Erika came under pressure from some of her co-authors on the original paper to clarify the position. Eventually, she conceded that the sequences were indeed wrong and, in August 2000, nearly eighteen months after the first paper appeared, the correction was published. For some unexplained reason, the sequences from the first part of the control region had been shifted by ten bases. This is something that can happen if the sequencing machine is playing up. The base that the machine had scored as a mutation at position 76 was actually the normal base for position 86. So there were no mutations at 76 after all. Getting to the truth had been an exhausting, unpleasant and distressing experience. Everyone makes mistakes. But to take so long to set the record straight on such an important issue with so many ramifications seems to me completely contrary to the spirit of scientific enquiry. But there it was. Mitochondria had survived the recombination scare.

12

CHEDDAR MAN SPEAKS

Although our scientific reasoning now appeared to be watertight, I was still nervous that there might be a flaw in our version of European prehistory that even our most persistent and vociferous critics had overlooked. They had done a good job in making us test and prove every conceivable aspect of our principal tool – mitochondrial DNA itself. We had checked and rechecked the mutation rate. We had spent weeks running different versions of our evolutionary network programs and they all gave the same results. We had ridden the storm of recombination. We still felt sure that main chapters of the genetic history of Europe were written in the time of the hunter–gatherers, long before the farmers arrived. To be sure, agriculture had added some important extra paragraphs; but it had definitely not erased the original text. We felt very confident that most living native Europeans traced their maternal ancestry back to the hunter–gatherers who lived before the dawn of the Neolithic and the coming of agriculture.

Nevertheless, even though we were very sure of our

data and the way we had interpreted them, our conclusions were still only inferences about past events:
inferences built on large amounts of data and robust
statistical treatments, but inferences none the less. So
I was still slightly anxious. Perhaps we had made a
mistake about the dates. I didn't think we had, but
suppose we were out by a factor of two? Suppose that
events we had dated to fifty thousand years ago actually
took place only twenty-five thousand years ago? More
importantly, suppose the dates for the major mitochondrial clusters which we had placed at or around
the end of the last Ice Age, between fifteen and twenty
thousand years ago, were out by the same factor and
were really less than ten thousand years old? That
would bring them too close for comfort to the Neolithic
period, and mean that they might have been part of the
wave of Near Eastern farmers after all.

What we needed was a direct test on DNA taken
from a human fossil which was known to predate the
arrival of farming. If we could only find DNA that
fitted into one of these crucial clusters in the remains of
a hunter who lived thousands of years before farming
was ever thought of, then we would be home and dry.
We would not need to rely exclusively on reconstructions from the modern sequences. We would have
found the real thing in Palaeolithic Europe. These
mitochondrial clusters then had to have arrived in
Europe thousands of years before farming ever reached
it, and our dates must be right. Conversely, if the DNA
from a very old fossil was unlike anything we now
found in Europe then we were on shaky ground. We

could not then be sure that the ancestors of the major modern clusters were in Europe before farming.

Human remains from the Upper Palaeolithic are few and far between. For one thing, ten thousand years is a very long time, and only in the very best of circumstances do bones last that long. Any that do survive are jealously guarded specimens, and rightly so. We would have to make an exceptionally good case to persuade a curator to allow us to take a sample from such a rarity. In my favour I did at least have a track record in getting DNA out of old bones. With my colleagues, I was the first to do so, with the Abingdon bones in 1989, although in that case the material was only a few hundred years old. Our work a few years later on the Iceman had become widely known, and was well thought of. But that was a unique case – a completely frozen body. At five thousand years it was old, but not old enough to predate agriculture. Although the Iceman's DNA belonged to one of the key clusters, it couldn't be used to strengthen our case because he was living two thousand years after farming had reached the Alps. We were looking for remains that were at least twice as old as the Iceman. Even so, he was the oldest human by far to have had his DNA successfully extracted, and as a deep-frozen body he was an exceptional case. There was no assurance that an ordinary skeleton would retain its DNA for five thousand years, let alone ten thousand.

Although DNA is obviously a much tougher molecule than anyone ever thought when they were scared to take it out of the refrigerator for fear of its decomposing, it cannot survive very long on its own. It needs to be in a skeleton to survive for thousands of years. What distinguishes bones, and teeth, from all other tissues is the hard, calcium-based mineral, hydroxyapatite. This protects the proteins and the DNA from decay by shutting out the bacteria and fungi that feed on the soft tissue in the rest of the corpse. So long as the mineral is intact, there is a chance that the DNA will have escaped being gobbled up. Once the calcium goes, the DNA is exposed and soon disappears. Calcium is alkaline and survives much better in an alkaline soil than anywhere else. In neutral and particularly in acid soils, DNA is much shorter-lived. The spectacular peat-bog bodies of northern Europe, where even the hair and skin are intact, always have a collapsed and deflated look about them because the calcium in the bones has dissolved in the acid bog. A lot of the protein survives and is protected against decay by the acid, which kills bacteria and fungi. However, because of its molecular structure, DNA is cut to shreds by even dilute acid very quickly. So, unfortunately, bog bodies are not a good source of ancient DNA.

Heat is also bad news. Egyptian mummies were an early, high-profile target for those in search of ancient DNA and, sure enough, some was found. But these were the carefully embalmed bodies of the wealthy, sheltered from decay not only by the natural preservatives in the embalming fluid but also by a succession of wood and stone sarcophagi which sealed the body in an underground tomb away from the baking heat of the sun. There are thousands of much less elaborate burials for the less well off in shallow graves just beneath the sand; but, even though these mummies are only two or three thousand years old, they are almost totally devoid of protein or DNA. The inorganic calcium is unaffected by the heat, but the organic molecules are long gone, broken down and leached away by the scorching heat of the desert.

We knew, then, that we had to avoid burials in hot countries and acid soils, and so we turned our attention to the limestone caves of northern Europe. Within these caves the temperature remains cool and, importantly, constant throughout the year. The daily fluctuations of heat and cold in the Egyptian desert probably do more damage to the DNA than the heat alone. A cool, stable temperature was much more promising. But what really recommends limestone is the alkaline nature of the surroundings. Bone mineral and lime-stone are chemically very similar. They are both compounds of calcium. The water that drips its way through the caves, forming stalactites and stalagmites, and covering the walls in sheets of flowstone, is rich in dissolved calcium. There is calcium everywhere. A bone left in a limestone cave does not have its mineral leached away. And if the mineral stays, and the temperature isn't too high, the DNA will stay as well.

The caves in Cheddar Gorge are the most famous in Britain. A small, winding road threads its way down from the top of the Mendip Hills about twenty miles west of Bath. At first it is like any other wooded valley in that part of the world. Ash and hawthorn trees flank the road and, in the spring, the woods are full of the white flowers and pungent smell of wild garlic. As you descend further, the sides of the valley get higher and higher and the trees retreat up the increasingly steep slopes until, only a couple of miles from the top, you are staring up at vast walls of limestone three hundred feet high. Except at the very bottom of the gorge there is no sign of the river which formed it. This disappeared underground long ago, where it dissolved caves and caverns out of the rock. As the roofs collapsed and collapsed again, so the gorge was formed. The newest caves are still there, not yet obliterated by the forces of water and gravity. In the bustling tourist town of Cheddar at the foot of the gorge, the caves are big business alongside the cheese for which the town is famous. On the left hand side of the gorge, directly opposite the Cheddar Caves Fish and Chicken Bar, and with its entrance partially obscured by the Explorer's Cafe-Bar and a shop, is the biggest and most spectacular cave of them all – Gough's Cave. And in the museum near the entrance to the cave stands a cast of its most celebrated former inhabitant: Cheddar Man. He was excavated in 1903 and subsequently carbon-dated to about nine thousand years ago, at least three thousand years before farming reached Britain. The cast is a copy of the original skeleton, which is stored in the Natural History Museum in London, in the care of Chris Stringer, head of the Human Origins Group. I rang him and made an appointment.

I knew Chris by reputation and had met him once at a scientific conference in Sardinia. The Natural History Museum I had known since my childhood. It was always a treat for my brother and me to be taken there by my mother in the school holidays. As I made my way up towards the immense and towering Victorian Romanesque entrance I felt a real excitement to be going to the Museum again not as a schoolboy but as a professional scientist. To reach Chris Stringer's office I had to walk past the skeleton of the huge dinosaur, *Diplodocus*, that dominates the magnificent entrance hall. Then I turned right into a wide corridor, its walls hung with the skeletons of Ichthyosaurs and other marine reptiles, still embedded in the blue clay of the Dorset cliffs where they were found. But when I went through the door into the palaeontology department, the atmosphere and the decor changed abruptly, from the dramatic to the professional. Row upon row of anonymous sliding cabinets concealed the treasures which lay catalogued within them. Chris Stringer's modern office led off from this priceless yet strangely silent testament to the wonders of the natural world.

Over a mug of tea, it didn't take long to explain my reason for wanting to sample human fossils from the Palaeolithic. He had read about the controversy which our work on European prehistory had sparked, and quickly saw the sense in testing the DNA from a pre-farming skeleton. He wanted to know what the chances were of our being able to recover any DNA if he were to give us permission to sample. I could not give a definite answer. After all, the Iceman was so unusual that I could not promise that because we had been successful with him we were guaranteed a good result with an unfrozen bone twice that age. Without that assurance Chris was understandably reluctant to give permission for us to take a destructive sample from something so precious as Cheddar Man. Remembering that we had also been successful with animal bones from the *Mary Rose*, I made a suggestion that I hoped would get us over this impasse. If there were any animal bones from Gough's Cave of approximately the same age, could we try them? If that worked, we could be fairly confident that the conditions within the cave were good enough to preserve DNA for ten thousand years. Happily, there were scores of animal remains from Gough's Cave and I went back to Oxford with a small piece of deer bone.

Within a month I was back in Chris's office with the good news. There was plenty of DNA in the deer bone. Chris agreed that this was sufficiently good proof to allow me to sample the human material. On the table in his office he carefully laid out the actual remains of Cheddar Man, each one enclosed in a cardboard box and supported by cotton wool. The skull had its own made-to-measure wooden case, with foam rubber supporting the delicate reconstruction from a dozen or more fragments cemented together. I didn't dare to touch it. Eventually we settled on the hallux, the solid-looking bone of the big toe. Chris packed it into a small cardboard box and I took it back to the lab.

Next day, I carefully drilled into the bone. What appeared from the outside to be solid bone was not. In no time I had punctured the thin shell of the cortex and was into the honey-combed interior. Black specks fell into the small pile of brownish bone powder from the drillings. These black bits certainly didn't look like bone; most likely they were bits of soil that had found their way into the middle of the toe-bone through a crack. I picked them out one by one with watchmaker's forceps and put them to one side. I had exactly 17.8 milligrams of Cheddar Man bone powder. It would just have to do; I didn't want to make another hole. By the following day I knew it was not going to work. There was no sign of any DNA. The control experiments had worked perfectly. Bright orange fluorescent spots, indicating the presence of amplified DNA, were in all the positive controls. The blanks, always run at the same time with water and not bone extract to control for contamination, were all blank. And so was the extract of Cheddar Man's toe. This was bitterly disappointing.

I went back up to London to talk things over with Chris. We knew from the success with the animal bone that the environment of Gough's Cave was good enough to preserve DNA for at least ten thousand years. Maybe the fact that the bones had been outside the cave for the best part of a century had something to do with it. Maybe the resin that was used to stabilize the bones had interfered with the DNA extraction. Or maybe there just wasn't any DNA there at all. Just so that we could have a focus for our thoughts as much as anything, Chris brought the skull back into his office and laid it out on his desk once more. I don't find it particularly easy to relate a skull to a living person but, as I looked at the pieces displayed on the desk, I began to imagine the flesh and the skin of the head building up on the reconstructed skull. As I write this it sounds distinctly macabre, but at the time it wasn't in the least. In my imagination, these were no longer just lifeless fragments of bone but a real person. I had no clear impression of what he looked like – no idea whether he had black or fair hair, brown eyes or blue – but I did have a very strong feeling that this was a person. Strange, remote, from a far-off time, but a person none the less. What stories he could tell about his life, his family. I picked up the lower jaw and looked at his teeth, the teeth he used to crush hazelnuts and tear into the flesh of freshly caught deer. The enamel was worn down, but the teeth were not rotten. In fact, they looked pretty healthy compared to my own set, which are full of fillings. When I idly mentioned this to Chris he turned and said, 'Well, if you think these are good, come and have a look at this.' He led me out of his office and into the large room with the storage cabinets. We walked to a distant part of the room and Chris brought out another small wooden box. He opened it and inside, nestling on its bed of foam rubber, was the lower jaw of a younger male. The teeth were absolutely perfect. White, regular and with no sign of decay. They could have come straight out of a toothpaste ad. I imagined they must be only a few hundred years old at the most. But they were not. These were the teeth of a young man who lived more than twelve thousand years ago – over three thousand years before Cheddar Man – and whom Chris had excavated himself from Gough's Cave in 1986.

Back in the brightly lit office, the teeth looked even

better. Could it be that, inside the teeth, the dentine and the pulp cavity would be much better protected even than the bone? Could the few molecules of DNA, which were all we needed to test our theory, be hiding inside the teeth encased in an unbroken shield of enamel? Even though we had failed with Cheddar Man's toe, we agreed it would be worth a shot. But no-one had any experience of extracting DNA from teeth, especially teeth still embedded in the jaw, and there was no question of being allowed to remove them to make it easier. I promised to go away and devise a method of drilling into teeth in a way that did not mark the enamel and allowed them to remain in the jaw. If I could do that, then Chris would allow me to take a sample from the Gough's Cave specimen.

I was back within a fortnight, having practised on some teeth given to me by my dentist, Mr Miller. I had perfected a way of drilling into a molar tooth and getting the dentine out with the tooth still embedded in the jaw, and I brought with me some samples of my handiwork for Chris to inspect. After trying and rejecting a straightforward dental drill (the compressed air blew the powder all over the place) I had found a small modelling drill which had been recommended by a colleague at work and which I bought from an ironmonger on the Tottenham Court Road in London. It was just perfect for making the small entry hole just below the enamel. Once I had got inside the tooth, another, longer drill bit could be attached and wiggled to and fro, reducing the soft dentine to a fine powder. I rigged up a suction device and, using this, it was very easy to remove the powder from inside the tooth into a small test tube. The hole then only needed to be filled with a colour-matching cement and the tooth looked as good as new – as it were. And the dentine, at least in my trial teeth, was full of DNA.

To avoid the ever-present possibility of contamination with modern DNA I needed to drill the teeth from the Cheddar fossils in my own laboratory, where we had recently installed a filtered-air clean room. We had bought it as a ready-made unit constructed for the silicon chip industry. The incoming air was filtered and maintained at a positive pressure, which meant that there was no chance of dust or flakes of skin getting into the room when you went in through the air lock. It was an expensive and elaborate precaution, but well worth it. So I had to take the jaw back with me to Oxford – which was a nightmare. I had come into London on the bus, and it was on the bus that I returned with this priceless and irreplaceable specimen in its box on the seat beside me. Every few seconds I would turn to make sure it was still there, trying to imagine what I could say if I lost it. Thank God, I didn't; and by late afternoon it was safely locked away in the specimen cabinet back in Oxford.

The next day I started the extraction. It couldn't have gone better. The drill sank into the second molar easily, but not too easily – that would have been a sign of bad preservation – and there was a slight smell of burning in the air. This was the collagen being vaporized by the speed of the drill, a smell I used to hate during my own visits to the dentist but one I had now come to love as a sign that there was plenty of protein left in the specimen – and where there is protein there is usually DNA. When I switched on the suction pump, the pale cream powder came flying out of the tooth into the tube. There was lots of it – just under 200 milligrams. I took 50 milligrams, so as to leave plenty for a repeat, and started up the extraction process.

By the following evening I knew I had mito-chondrial DNA from the tooth. Over the next two weeks I read through the sequence, checked it again and confirmed it by a second extraction. I was looking at the DNA sequence of the oldest human fossil, by far, that had ever been successfully extracted anywhere in the world. But that wasn't the most important thing. The crucial piece of information we were looking for was embedded in the details of the DNA sequence itself. Was this the same sequence as a thoroughly modern European, or was it an obscure relic that was now extinct?

The answer was crystal clear. The ancient DNA from Gough's Cave was also completely modern. The sequence lay at the centre of the largest of the seven mitochondrial clusters. It is by far the commonest sequence in modern Europe; and here we had found it in the tooth of a young man who had lived fully seven thousand years before the arrival of farming in Britain. Here was the proof that this sequence, this cluster, and, by extension, the others of a similar estimated age were well and truly established in Europe long before the farmers. The Upper Palaeolithic gene pool had not been fatally diluted by the Middle Eastern farmers.

There was more of the hunter in us than anyone had thought.

Though I had got no further than drilling into his big toe, this was not the last encounter I had with Cheddar Man. We were re-introduced, so to speak, as part of a television documentary. Philip Priestley, an independent producer, was setting up a series of archaeology-based programmes for a west country TV station, and one of them was built around the excavation of a Saxon palace in Cheddar. By now our work on the genetic continuity between the Palaeolithic and the present day was reasonably well known, and it occurred to Philip that it would make good television if he could relate, through DNA, some of the present-day residents of the town with Cheddar Man himself. This seemed both fun and worthwhile; but I explained that we had already had a go at getting DNA from the Cheddar Man remains without success. If he could get permission from Chris Stringer, I was willing to try again, this time with the teeth, not the toe-bone, but only on condition that if nothing came of it we would not be filmed. I always work on that basis. I have seen too many programmes that begin with a big build-up anticipating a great scientific discovery at the finale, only to peter out in an inconclusive or unsuccessful experiment. So, with all the ground rules agreed, and after another nerve-racking journey on the bus, this time with an even more famous fossil in a box beside me, I drilled into Cheddar Man's first molar.

Out came the powder – not quite as clean as the earlier Gough's Cave material, but in sufficient quantity for an extraction. We found enough DNA for a reasonable sequence and were not surprised when it fitted comfortably into one of the seven clusters. Philip, understandably nervous as the deadline for filming got closer, was delighted and immediately organized the second strand of the piece, the sampling of the Cheddar residents. The site of the Saxon palace, featured in another programme in the series, is in the grounds of the local secondary school, and it made good sense to approach the school to see if they would agree to their pupils taking part in the programme. By now we had refined our DNA sampling procedure. We no longer used blood samples; instead we found that a small brush rubbed gently against the inside of the cheek picks up enough cells from the surface to give us plenty of DNA. After a short visit to the school, we had twenty samples from the sixth form volunteers and some of the teachers. Knowing how often we had found Cheddar Man's sequence in modern Britain, I reckoned there was a fifty–fifty chance of getting a close match in the twenty samples we had taken. Within four days we had the results. We knew the names, and (crucially, as it turned out) the ages of the volunteers. Philip was on the phone.

'We've got a match,' I told him.

'Who is it?' was his first question.

This wasn't part of the deal. While we had agreed to see if we could find a match among the twenty residents, I had not agreed to identify any individuals, for a very good reason. Although the children, and their parents, had signed forms consenting to have their

DNA sampled and to take part in the television pro-gramme, I felt there was a risk that they might not have realized what they were letting themselves in for if the story broke in a big way. Though there is no way of knowing beforehand how big a story is going to become, the experience of Marie Moseley and the Ice-man was an indication of its potential.

At this point Philip became distinctly agitated. He thought the story would be worthless without an individual identification. He immediately faxed me a copy of the consent form, but as far as I could see it was just a standard release – not, in my opinion, sufficient as a basis on which to claim consent to a possible worldwide media intrusion into the life of a teenager. I checked our list of sequences against the names and ages of the volunteers. There was not one match but three: two exact matches with Cheddar Man, and one with a single mutation; and while the two exact matches were children, the close match was a teacher, in fact the head of history who was organizing the filming in the school, Adrian Targett. I made the decision that I would identify Adrian Targett but not the two children. As it turned out, it was one of the best decisions I ever made. Unknown to me, Philip and his publicity team had organized a public 'reveal' where Adrian Targett would be identified as Cheddar Man's relative in front of the cameras and in the presence of a television news crew. They, too, were beginning to sense the potential magnitude of the story. The next day, when I went to the newsagent, I could not believe my eyes. The story of Adrian Targett and Cheddar

Man was in *all* the papers: from the London *Times* to the tabloid *Daily Star*, there was Adrian on the front page, posing beside his famous fossil relative. I bought the lot.

In the following days and weeks the story of Cheddar Man spread around the world. I met Adrian Targett on a TV chat show. He told me how one tabloid newspaper, famous for its pictures of topless women, had offered him a five-figure sum (so at least £10,000) to pose in a fur loincloth beside his ancient relative. Being a sensible man, conscious of his standing as a teacher, he declined. But it did make me wonder what the newspaper would have offered a teenage girl to wear the same outfit – or less. Even now, years later, people still remember the Cheddar Man story, if not always accurately. I was talking to an American audience in 2000 on something completely different when a woman asked me: 'Are you the one who did the DNA from the Cheese Man?' At the time, not surprisingly, I had a full postbag for weeks after the story broke. Many letters were complimentary, including a very well-informed one from the inmates of San Quentin gaol in California, who were keen to discuss the findings at the next meeting of their anthropology study group. But the one that stood out came from the secretary to Lord Bath. It turned out that Cheddar Caves are part of Lord Bath's estate. Evidently he had read the story (though whether in the *Times* or the *Daily Star* I never discovered) and wanted to know if he too was related to Cheddar Man.

Alexander Thynn, Lord Bath, is the owner of

Longleat, one of the most beautiful houses in England. It is famous for the safari park in the grounds, where visitors can watch the famous Longleat lions and other dangerous animals from the alleged safety of a car. Lord Bath himself, affectionately referred to as the Loins of Longleat, is well known for his idiosyncratic personal life. In addition to a legitimate wife and children, he has a stable of what he calls his 'wifelets', many of whom live on the estate. This was definitely one to follow up, and the next weekend I was on the way to Wiltshire. I was led upstairs to the penthouse suite on the top floor of this magnificent Elizabethan house. Lord Bath himself, now in his sixties but with a youthful twinkle in his eye, was dressed in one of his collection of brightly coloured kaftans that bulged from a wardrobe close to an absolutely enormous wooden desk. The life clearly suited him. He poured out two large glasses of rosé from a tap on the wall as I went through the genetics with him. A few glasses later we got round to the test itself, and he brushed the inside of his cheek. During the course of the morning several other people passed through the penthouse, and each was encouraged to give a sample, which they cheerfully did. He was evidently very popular with his staff. By lunchtime we had half a dozen DNA brushes and it was time for me to leave.

When we got the results back it came as no surprise that Lord Bath was not closely related to Cheddar Man. There was no particular reason why he should be. But his butler, Cuthbert, one of the other people who had given a sample during my visit to Longleat, *was* an exact match. At a stroke he could claim an ancestry which stretched back nine thousand years, making the five-hundred-year pedigree of the Thynns look distinctly *nouveau*. I asked Lord Bath how Cuthbert had received this piece of news. Had it made him reassess his attitude to the aristocracy? 'Well,' he replied with a smile, 'he has been feeling very confident lately.'

We had now done about as much as we could to establish our claim that the maternal ancestors of the majority of modern Europeans were already living in Europe well before the arrival of farming. We could not say anything about other genes, only about mito-chondrial DNA; but on this basis we had a clear picture of European prehistory, built up from both modern and fossil DNA, not of a massive replacement of the hunter–gatherers by the farmers but of a strong con-tinuity back to the days of the Palaeolithic. There was only one of Cavalli-Sforza's criticisms that we could not answer. Whatever way you look at it, mitochondrial DNA is only one gene and, as such, subject to statistical fluctuations that might make it unrepresentative of the human genetic legacy as a whole. I did not think this very likely; but what was needed to substantiate our version of European prehistory was confirmation from another gene altogether.

13

ADAM JOINS THE PARTY

The story I have narrated in this book is a history of the world recorded in the gene that is the easiest to read, mitochondrial DNA. So far, then, it is the gospel according to Eve. The beauty and simplicity of viewing the record of the past through mitochondrial DNA derive from its unique genetics, and in particular from the clear message that passes virtually unchanged from generation to generation, modified only by the slow ticking of the molecular clock as mutations gradually build up one at a time.

It would be strange indeed if a second, completely different, history were to be encrypted in the other genes that we carry. All these other genes are found on the chromosomes of the cell nucleus. According to the latest estimates, there are just under 30,000 of them. Are there 30,000 different versions of the human past waiting to be read? In one sense there are, because each of these genes could have a different history. Each of them might have a different common ancestor some-where in the course of human evolution. However, while our nuclear genes have percolated down through time, it is quite impossible to trace all these lines back along a known pathway of descent in the way that we were able to do with mitochondrial DNA. The reason is that, unlike mitochondrial DNA, the nuclear genes are inherited equally from both parents. While you have only one mitochondrial ancestor in the last generation, your mother, you have two nuclear ancestors, your mother and your father. That doesn't sound too complicated. But go back one more generation. Now you have four nuclear ancestors, your grandparents; but still only one mitochondrial ancestor, your mother's mother. Go back another generation and there are eight nuclear ancestors, your great-grandparents; yet *still* only a single mitochondrial ancestor, your grandmother's mother. At each generation the number of nuclear ancestors doubles. Go back twenty generations, to about AD 1500, and there could be, theoretically, over one million ancestors who could have contributed to your nuclear genes. In practice, many of these potential ancestors will actually be the same individuals, whose lines of descent have come down to you along different pathways, crossing between males and females through the generations in an unpredictable way.

Tracing the genealogy of all 30,000 genes through this maze of interconnections would be quite impossible. Add to that the confusion introduced by recombination, and the magnitude of the task becomes mind-numbing. The shuffling of chromosomes at each generation means that any one gene might itself be a combination of one part from one ancestor and another from someone else. Reading the different individual versions of human history from these genes, and bits of genes, in the cell nucleus is impossibly complicated at the moment. It will take a long time to advance beyond the kind of crude summaries of human history that we already have from the days of gene frequency comparisons.

However, one gene – or, more correctly, one chromosome – is immune from these ghastly complications. It is called the Y-chromosome, and it has only one purpose in life: to create men. By comparison with the other human chromosomes it is small and stunted, and it carries only one gene which really matters. This is the gene that stops all human embryos from turning into little girls. Without a Y-chromosome, the natural course of events is for the human embryo to develop into a female. If an embryo has a Y-chromosome, and if the gene, which has been given the undistinguished name SRY, is working properly, then it will trigger a number of other genes on different chromosomes to steer the development of the embryo away from becoming a female and towards becoming a male. The SRY gene activates genes on other chromosomes which suppress the development of ovaries and instead promote the growth of testes and the production of the male hormone testosterone.

Two observations pinpointed the key part played by the SRY gene in sex determination. Very rarely, in something like one in 20,000 births, a girl is born with a Y-chromosome. These girls look normal, they have normal intelligence and they develop normally, though they are usually slightly taller than average. But at puberty their ovaries and uterus do not develop properly, and they cannot have children. Genetic analysis of the Y-chromosomes of these girls shows that the SRY gene is either missing altogether or contains a mutation that stops it working properly. The other piece of graphic evidence that the SRY gene is itself sufficient to make a male came from research on mice. Male mice have Y-chromosomes too, and they carry the mouse equivalent of the human SRY gene – called, in a burst of imaginative classification, Sry. In a very elegant genetic engineering experiment, the Sry gene was cloned from a male mouse and transplanted into a fertilized mouse egg that would otherwise have turned into a female. Despite the fact that the mouse embryo had only the cloned gene to work on, rather than a complete Y-chromosome, it turned into a male.

So this is how the sex of a baby is determined. Fathers, being male, have a Y-chromosome. Half of their sperm contains this Y-chromosome, carrying the SRY gene, and the other half carries another chromosome – the X-chromosome – instead. The sex of the baby depends entirely on whether or not the particular sperm that fertilizes the mother's egg contains an X- or a Y-chromosome. If the successful sperm carries an X-chromosome, then the child will be a girl. If it carries a Y-chromosome instead, the child will be a boy. The woman has no influence whatsoever on the sex of the child. How many women in past centuries would have loved to know this simple fact? How often was the 'failure' to produce sons attributed to a failure, deliberate or not, on the part of wives to conceive boys?

Just as mitochondrial DNA follows a maternal genealogy through the generations, the inheritance of Y-chromosomes by sons from their fathers should trace the mirror-image paternal pathway from one generation to the next. If the Y-chromosome could be genetically typed, and if it were not involved in recombination that would scramble the message, then there was good reason to believe that it would be the perfect complement to mitochondrial DNA in reading the history, not of women, but of men. The Y-chromosome, in common with all the chromosomes of the nucleus, is a very long, linear molecule of DNA. While mitochondrial DNA has just over sixteen and a half thousand bases in its DNA circle, the Y-chromosome stretches for about sixty million bases from one end to the other. It might be the runt among human chromosomes, but it still packs more than four thousand times as much DNA as mitochondria. Moreover, there is some gene shuffling within it. At the tips of each end of the Y-chromosome there is a section of DNA that recombines with the X-chromosome; but since these sections involve less than 10 per cent of the whole chromosome, this doesn't present a great problem. Genes that are on the recombining part of the Y-chromosome will trace a mixed genealogy, swapping unpredictably from males to females just like all the other nuclear genes. However, the remaining 90 per cent of the Y-chromosome, between the recombining ends, is not scrambled. This long segment travels intact through the generations. But are Y-chromosomes different from one another, and if so how do they differ? Only if there were variety and diversity in the Y-chromosome would it be of any value at all for reading human history. If all Y-chromosomes were exactly the same, they would be no use for our purposes.

Chromosomes are intensively studied under the microscope by trained cytogeneticists in medical genetics laboratories who are on the lookout for abnormalities that can diagnose inherited diseases like Down's syndrome or explain the cause of infertility. With all this activity going on, cytogeneticists had noticed that some Y-chromosomes stood out as being much longer than the average. This was promising; but it was not a very precise way of differentiating between Y-chromosomes on a large scale. Besides, the lengths were unstable and changed between one generation and the next. What was needed was the same kind of testing involving Y-chromosome DNA that had identified mitochondrial DNA as such a star. Then it would be straightforward to fingerprint Y-chromosomes from hundreds or thousands of volunteers easily and cheaply. But how were the segments of Y-chromosomes that were going to show the biggest differences among people to be found?

The rich diversity of the mitochondria is concentrated in a small DNA circle of only a few thousand bases. Better still, the control region compresses about a third of the diversity of the whole mitochondria into just five hundred bases that can be sequenced in a single run on an automated sequencing machine. Would something similar be found in the

Y-chromosome? The answer was not long in coming. Several labs, hoping for the best, began to look for differences between Y-chromosomes by sequencing through the same segment of Y-chromosome DNA from volunteers who were as distantly related to one another as possible. In one of the first studies, 14,000 bases were sequenced from the Y-chromosomes of twelve men from widely different geographical origins. Only a single mutation was ever found. If an equivalent 14,000 bases had been taken from mitochondrial DNA instead of the Y-chromosome, they would have shown dozens of mutations in the same number of people. Another lab sequenced a 700 base segment of one gene from the Y-chromosomes of thirty-eight different men without finding a single difference in any of them!

This was all rather depressing for the scientists involved (thankfully, I wasn't one of them). There was a lot of head-scratching. Why were Y-chromosomes so similar all around the world? Since Y-chromosomes didn't carry any genes to speak of, and were full of 'junk' DNA which had no obvious function, the expectation was that there should be more, not less, variation on the Y-chromosome than on regular, gene-rich chromosomes. Mutations are free to accumulate in 'junk' DNA because it doesn't do anything, so its precise sequence doesn't really matter. Most mutations that occur in genes which do have important functions interfere with their proper working and are soon eliminated by natural selection. It was certainly a puzzle to find that there were so few mutations on the Y-chromosome.

The most popular theory advanced to account for this lack of variation was that it had to do with the fact that, under the right circumstances, men can have a lot more children than women. If, in the past, only a few men had lots of children, and therefore lots of sons, their Y-chromosomes would spread around quickly at the expense of the Y-chromosomes of their unfortunate male contemporaries who had few children or none at all. If this had happened a lot, the theory went, there would be far fewer different Y-chromosomes around today than if all men had roughly the same number of children. It's true that there have been some particularly prolific males. The world record holder is Moulay Ismail, Emperor of Morocco, who is alleged to have had 700 sons (so presumably as many daughters) by the time he was forty-nine in 1721. He died in 1727 – so there was another six years to have some more. The most prolific woman comes way behind this. She is Mrs Feodora Vassilyev, a Russian woman who produced sixty-nine children between 1725 and 1765. They were all multiple births – sixteen pairs of twins, seven sets of triplets and four lots of quadruplets – so she was a remarkable woman in that respect as well. The capacity of women to produce large numbers of children is limited by their biology, which restricts them to one pregnancy a year at most. Men, on the other hand, are not restricted by this timetable and can, in theory, have thousands of children. But the fantasy of enormously prolific males seeding the entire world, thereby reducing the diversity of Y-chromosomes by their prodigious feats of polygamy, turned out to be just that. A fantasy. A hard slog in laboratories around the world over the past ten years has found that there are plenty of mutations on the Y-chromosome after all.

These mutations come in two main types. The first is exactly the same as those we are already used to seeing in mitochondrial DNA: the simple change from one base to another. However, unlike in mitochondria, where they are neatly compressed into the control region, these mutations are spaced out at irregular intervals right along the length of the Y-chromosome. This is a practical nuisance because each one has to be tested individually, but it is not an insuperable obstacle. The other type of mutation is very uncommon in mitochondria, although we did encounter one example in the Polynesian samples. That is where there was a deletion of nine bases from the mitochondrial DNA circle. A careful look at the DNA sequence around that region revealed that in fact this wasn't so much a deletion from the Polynesian mitochondrial DNA as a doubling up, a duplication, of that nine-base segment in the rest of us. This type of mutation, where short segments of DNA are repeated over and over again, is remarkably common in the nuclear chromosomes and, thank heavens, in this respect the Y-chromosome is no exception. Dozens of these repeated segments have been discovered on the Y-chromosome, and the difference between individuals lies in the number of repeats. Fortunately, this is an easy thing to measure. This rich source of variation suddenly revealed that there are thousands of different Y-chromosomes around that can be distinguished from one another on the basis of these two sorts of mutation. Genetic finger-printing of Y-chromosomes has become a reality.

Because it has been such a struggle for the scientists involved to find the useful mutations, laboratories have been very careful about whom they tell when they find a new one. As a consequence, labs have organized themselves into rival cliques which have used different sets of mutations to fingerprint Y-chromosomes; there is not yet a common standard. This means that there are different evolutionary networks being produced by the different confederations of laboratories. This is only a temporary situation, and I hope and expect that in the near future these will be reconciled into a scheme which everyone can accept. But how is it looking up to now? In particular, does the history of Europe revealed by the Y-chromosome bear any resemblance to the one read from mitochondrial DNA which forms the basis for this book? Does the Y-chromosome version of events agree or disagree with the mitochondrial DNA in placing such a great emphasis on the Palaeolithic as the source of our genetic legacy? In other words, does the history of men agree with the history of women? The answer came in an article published in the 10 November 2000 edition of the journal *Science*.

'The genetic legacy of Paleolithic *Homo sapiens* in extant Europeans: a Y-chromosome perspective' was the culmination of a large collaboration between scientists from Italy, eastern Europe and the United States. I had been asked to comment on the paper by the BBC on the day it was published, and had a copy faxed through to the Royal Society in London where I was at a scientific meeting. As soon as the fax arrived I took it into one of the drawing rooms which overlooked St James's Park and sat down. My heart sank as I went through the long list of authors at the beginning of the paper. There, second from the end, was the name L. Luca Cavalli-Sforza. After all the battles of the previous four years, I could hardly expect my old adversary to agree with me at last.

Reading through the article, I could see that it was constructed along generally similar lines to our mito-chondrial paper of 1996. They had fingerprinted the Y-chromosomes of 1,007 males from twenty-five European and Middle Eastern locations. Then, just as we had, they had drawn an evolutionary frame-work and identified clusters. They discovered ten Y-chromosome clusters rather than the seven that we had found with mitochondria. Then they had estimated the ages of these clusters, as we had done, from the accumulated mutations within each one. I turned the pages with growing excitement. What were the ages of these clusters going to be? Would they be mostly in the Palaeolithic, like six of the seven mitochondrial clusters? Or would they be much more recent, in the time of the Neolithic and the early farmers? I certainly knew what I expected the paper to say, given Luca's prominent position as an author and his well-known views on the magnitude of the genetic impact of agriculture. The paper was full of dense statistics but there, on the penultimate page, my eye went straight to the vital paragraph. It began: 'Analyses of mito-chondrial DNA sequence variation in European populations have been conducted,' and it referenced our 1996 paper. 'These data suggested', it continued, 'that the gene pool has about 80% Palaeolithic and 20% Neolithic ancestry.' That was fair. I read on to the next sentence, expecting it to begin the demolition of our position. But, it did not. Instead, I read the words: 'Our data support this conclusion.'

I couldn't believe it. The tension drained from my body. The battle was over. We had been put through the wringer for four and a half years. We had endured the panics about the mutation rate being wrong, about mitochondrial recombination messing everything up, and about the control region being completely un-reliable. And now it was over. Mitochondrial DNA and the Y-chromosome told the same story. The history of men tallied with the history of women. Luca and I could finally agree. It had been a tough battle, but a fair one. The Neolithic farmers had certainly been important; but they had only contributed about one fifth of our genes. It was the hunters of the Palaeolithic that had created the main body of the modern European gene pool.

14

THE SEVEN DAUGHTERS

From the remains in Cheddar Gorge we had extracted direct proof of the genetic continuity between people living today and the hunters of the Upper Palaeolithic. We now knew that this unbroken thread, accurately and faithfully recorded in our DNA, stretched back beyond the beginnings of history, beyond the ages of iron, bronze and copper to an ancient world of ice, forest and tundra. Only the exceedingly slow beat of the molecular clock separated the DNA we found in Cheddar Man from the DNA in our two utterly modern descendants Adrian Targett and Cuthbert the butler. The evolutionary reconstruction we had done on the DNA from thousands of living Europeans had pointed us to that conclusion, and eventually we had found physical evidence to validate it. Now we also had the crucial endorsement from another genetic system altogether, the Y-chromosome, of the assertion that our genetic roots do indeed go back deep into the Palaeolithic.

Our reconstructions had identified seven major genetic clusters among the Europeans. Within each of these clusters, the DNA sequences were either identical or very similar to one another. Over 95 per cent of modern-day native Europeans fit into one or other of these seven groups. Our interpretation of European prehistory and the emphasis it placed on the Palaeolithic hunter–gatherers had depended on giving ages to these clusters, and we had worked these out by averaging the number of mutations we found in all the modern members of the seven different clans. This gave us a measure of how many times the molecular clock had chimed within each clan. Knowing the rate at which the clock ticked, we could then work out how old each clan really was. Old clusters had accumulated more changes over the millennia. The molecular clock, slow as it is, would have struck more often. Young clusters, on the other hand, would not have had as much time to accumulate as many changes, and the DNA sequences of people within a young cluster would be more alike.

The seven clusters had ages of between 45,000 and 10,000 years. What these estimates actually tell us is the length of time it has taken for all the mutations that we see within a cluster to have arisen from a single founder sequence. And, by purely logical deduction, the inescapable but breathtaking conclusion is that the single founder sequence at the root of each of the seven clusters was carried by *just one woman* in each case. So the ages we had given to each of the clusters became the times in the past when these seven women, the clan mothers, actually lived. It required only that I gave them names to bring them to life and to arouse in me, and everyone who has heard about them, an intense curiosity about their lives. Ursula, Xenia, Helena, Velda, Tara, Katrine and Jasmine became real people. I chose names that began with the letter by which the clusters had been known since we had adopted Antonio Torroni's alphabetic classification system. Ursula was the clan mother of cluster U. Cluster H had Helena at its root. Jasmine was the common ancestor for cluster J; and so on. These were no longer theoretical concepts, obscured by statistics and computer algorithms; they were now real women. But what were they like, these women to whom almost everyone in Europe is connected by an unbroken, almost umbilical thread reaching back into the deep past?

There are a few qualifications you needed to be a clan mother. The first is that you needed to have daughters. That is obvious, because the gene we are following, mitochondrial DNA, is passed from mother to daughter. A woman who had only sons could not be a clan mother because her children would never pass on the mitochondrial DNA they received from her. So that is the first rule. The second is that you had to have at least *two* daughters. It's easiest to see why by looking at things the other way round, from the present to the past. The clan mother is the *most recent* maternal ancestor that all the members of a clan have in common. Imagine a clan with ten million living members and imagine that we knew perfectly from the registry of births, marriages and deaths exactly how they are all related. As we went back in time, generation by generation, we would see the maternal lines slowly joining up. The lines in brothers and sisters would converge, after just one generation, in their mother. After two generations, cousins would converge on their maternal grandmother, their mother's mother. Three generations ago it would be the second cousins whose lines coalesced in their maternal great-grandmothers. And so on. At each generation there would be fewer and fewer people in the clan who had maternal descendants living today. Eventually, hundreds or even thousands of generations ago, there would be only two women in the clan who could claim to have maternal descendants living in the twenty-first century. Further back still, the maternal lines of these two women would converge on a single woman – the true clan mother. And to be in that position she must have had not one but *two* daughters.

To clarify this rather tricky point take a look at Figure 5. I have drawn out an imaginary maternal genealogy of fifteen living women, represented by the white circles on the right-hand side. Only the ancestor marked by the arrow is the *most recent* common ancestor of all fifteen. Her mother is also a maternal ancestor of all the women, but she is not the most recent. Her daughter is. Equally, *her* two daughters, marked with asterisks, are both maternal ancestors of living women, but neither daughter is the maternal ancestor of all fifteen of them. If we called this a clan, then only the woman with the arrow is the clan mother. Exactly the same principle applies whether there are fifteen people in a clan, or fifteen thousand or fifteen million. There is still only one clan mother.

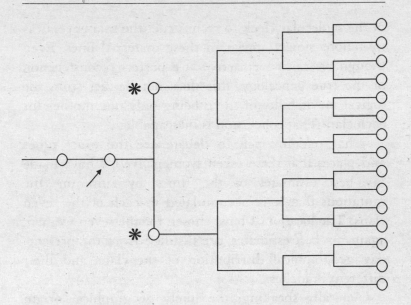

Figure 5

A clan mother did not have to be the only woman around at the time and she certainly wouldn't have been. But she is the only one who is connected through this unbroken maternal thread right through to the present day. Her contemporaries, many of whom will themselves have had daughters and grand-daughters, are not clan mothers because at some point between then and now their descendants in the female line either had no children or produced only sons. The lines died out. Of course, since we do not have records going back more than a few hundred years, let alone a few thousand, we can never hope to know the precise genealogy all the way back to the clan mother. All we can do is use the DNA sequences and the slow ticking of the molecular clock to reconstruct the main events as mutations slowly appear in these maternal lines. Even though we can never arrive at a perfect reconstruction of the true genealogy, this does not detract from the logical inevitability of there being only one mother for each clan. That conclusion is inescapable.

What remain open to debate are the exact times and places that these seven women lived. I have made my best estimates of the times by summing the mutations that have accumulated in each of the seven clans. The locations I have chosen for the seven women, again my best estimates, are distilled from the present-day geographical distribution of the clans and their different branches.

Generally speaking, the likely geographical origin for a clan is not necessarily the place where it is most common today but the place where it is the most varied. For example, going back to the Pacific, the clan that is very common in Polynesia did not originate there. Even though it is extremely abundant, there is very little diversity within the clan in Polynesia: most Polynesians who are in that clan today have the same DNA sequence. On genetic grounds alone, the origin of the clan is much more likely to be further west in the islands of Indonesia around the Moluccas. Even though the clan is not particularly common on the Moluccas today, there is a lot more variation within it there than in Polynesia. Only a fraction of the population moved out to Polynesia, so the diversity within the clan drops. In native Taiwanese, the diversity within the clan is even higher although, as in the Moluccas, it is not especially common. That makes it likely that Taiwan is an even earlier origin of the Polynesian clan than the Moluccas. When it comes to Europe, although we lose the simplicity that comes through dealing with discrete island populations, the same considerations apply. Clan origins are likely to be near the locations where they are the most variable today. Even so, this somewhat theoretical argument has to be tempered with realism. The mother of a clan which is twenty thousand years old cannot have lived in the north of Scotland, even if that might be where the clan is most varied today, for the very practical reason that Scotland was covered in ice at the time. I freely admit that there is a considerable element of uncertainty about exactly where these women lived. Indeed, while I would be alarmed if an equal uncertainty surrounded the exacting science behind the genetics, I somehow feel an element of mystery surrounding certain aspects of these seven individuals is not inappropriate.

As I became more engrossed in these seven women, I began to imagine what existence was really like for them. I was filled with an intense curiosity about their lives. Having let the genetics direct me to the times and places where the seven clan mothers most likely lived, I drew on well-established archaeological and climatic records to inform myself about them. The record of past temperatures is held in the frozen cores taken from the polar ice caps. Raised and submerged beaches mark out the sea-level changes which have been such a feature of the past fifty thousand years. The vegetation leaves its mark in pollen which survives for thousands of years after it was shed by the flower that made it. The changing styles of tools made from stone and bone that are excavated from sites of human habitation record the ebb and flow of technological progress. The animal and fish bones that litter the same sites tell of our ancestors' diet. All these pieces of tangible evidence combine with the genetics to recreate the imagined lives of these seven women, Ursula, Xenia, Helena, Velda, Tara, Katrine and Jasmine. They were real people, genetically almost identical to us, their descendants, but living in very different circumstances. What lives they must have led.

Come with me now on a journey into the deep past. Guided by the unbroken genetic threads that link us to our ancestors, we can travel back to a time before the dawn of history, to a world of ice and snow, of bare mountains and endless plains, to meet these remarkable women – the Seven Daughters of Eve.

15

URSULA

Ursula was born into a world very different from our own. Forty-five thousand years ago it was a lot colder than it is today, and would get colder still in the millennia to come leading up to the Great Ice Age. Ursula was born in a shallow cave cut into the cliffs at the foot of what is now Mount Parnassus, close to what was to become the ancient Greek classical site of Delphi. The cave mouth looked out across a wide plain a thousand feet below which led away to the sea twenty miles off to the south. Today this same plain is filled with the dark green of ancient olive groves; then it was a landscape of scattered woodland pressed up close against the mountain slopes with open grassland beyond. The coastline was several miles further from the cave than it is today. This was a consequence of the lower sea level that prevailed when more of the oceans' water was locked into the ice and snow of the polar ice caps and enormous glaciers filled the valleys of the great mountain ranges. Temperatures would carry on falling for another twenty-five thousand years as part of the regular climatic cycle that has been going on for at least four hundred thousand years and will no doubt continue far into the future.

Of course, Ursula was completely oblivious to these long-term changes – much as we are today in our everyday lives. What mattered to her and her band of twenty-five was the here and now. Ursula was her mother's second child. The first had been taken by a leopard when he was only two, in a raid on a temporary camp one dark night. This was a tragic but not uncommon occurrence in Ursula's world. Many children, and occasionally adults too, were hunted and killed for food by lions, leopards and hyenas. Though it was a sad and serious blow for Ursula's mother to lose her only child, it did at least mean she could get pregnant again. While she was nursing her son her periods had stopped, she no longer ovulated and could not conceive. This was a deliberate evolutionary adaptation to space out the children. Only when one child could walk well enough to keep pace with the seasonal migrations of the band would another be conceived. And that could take three or even four years. So, a year after she lost her son, she gave birth to Ursula.

It was March, the days were getting longer and the band had moved up from the coast where they had spent the winter. It was a good time of year; Ursula's mother always looked forward to the spring. The coast in winter was damp and miserable. There were no caves to shelter in and she had to do the best she could in crude shelters of wood and animal skins. It wasn't much of a home, and the living was difficult and uncomfortable to say the least. But they had to come down from the mountains: it was too cold up there, and in any case all the game on which they depended had retreated to the lower ground. There was plenty of it, but it was hard to catch. Her particular favourite was bison, which congregated on the plain in reasonable numbers at that time of year. But they were practically impossible to hunt down on foot and in the open. It was difficult and dangerous work. They were wary, hungry themselves and very bad-tempered. Only the year before two young men had been trampled to death in a stampede; since then, everyone had decided that it was just not worth it, and bison hunting in the winter was off limits. The loss of two hunters from the small band was a serious business, because it meant that there were extra mouths to feed in the shape of the bereaved women and their children. But the band only survived by co-operation, and there was no question of abandoning the dependants to their fate.

With bison hunting ruled out, the only food coming into the winter camp was either scavenged from carcasses or the occasional red deer that could be ambushed in the woods higher up the slopes. Scavenging was a depressing business for the hunters, and not without risk. They walked for miles, keeping an eye open for the signs of a kill made by a lion or a leopard. They might be lucky and spot the kites circling over-head if it was a clear day, but more often than not it was just a question of trudging round the usual circuit and listening out for the dreadful chatter of hyenas as they fought over the rapidly disappearing carcass. There had to be at least five people for a successful raid against a pack of hyenas. Making as much noise as possible, they rushed the carcass and scattered the hyenas before the beasts had a chance to realize what was going on. Then two of the group got on with the business of slicing off whatever meat was left while the others confronted the caterwauling hyenas that always hung around and made repeated lunges at either people or carcass. They pelted the beasts with stones and yelled to keep them back until the butchers had salvaged what they could, including the ribs, which were rich in marrow. Then it was a question of a hasty and organized retreat, with more stone-throwing and shouting as they left the scene. The trick was to leave at least some of the carcass behind, and to cover up what they had managed to collect under a skin. That way the hyenas eventually stopped following and returned to what was left. It was miserable, degrading work. The hyenas were awful, with saliva dripping from their disgusting mouths and making that frightful noise. There was nothing noble about this way of making a living, and everyone wanted to get off the soggy plains as quickly as possible and back to the mountains where at least they could hunt properly.

As soon as the first swifts appeared overhead, back from spending the winter in Africa, the band struck camp and started north for the mountains. The idea was to get there before the bison moved up to their summer pastures on Parnassus; that way there was a good chance of ambushing them as they filed through the steep-sided gorge below the cave. But even that wasn't simple. If men had been trampled in a bison stampede on the open plain, imagine how much more dangerous the herd was in the tight confines of a gorge only 10 metres across at its narrowest point. As usual, there was an argument about the best way to go about it. This happened every time. Some people advocated blocking the gorge and diverting the lead animals into a side couloir where they could be stoned and speared to death. The trouble with this approach was that some bison, who definitely sensed what was going on, had a nasty habit of turning round when cornered and charging straight back. The prospect of facing a ton of charging muscle and horn was too much for some people, and they shot up the rock face. When the escaping animal got back, snorting and sweating, to the main herd, this panicked the whole lot and they charged through the gorge at enormous speed. The advocates of a less audacious method pointed out the dangers of this direct attack and argued that it was simpler to wait until the main herd had gone through the gorge and pick off the stragglers. This wasn't a particularly heroic approach, but it did usually work. The bison bringing up the rear were usually the old members of the herd, but they still tasted better than scraps scavenged from the hyenas.

While this argument was going on, Ursula's mother withdrew to the shelter of the spring camp in the cave. Even though it was not uncommon for children to be born when the band was on the move, it was a lot more comfortable in a settled camp. The cave was dry and it was warming up as the sun rose higher in the sky. She was very glad to have made it before the birth. From the smell that hung around the back of the shelter it was obvious that it had been used as a winter lair by a cave bear. These huge and fearsome creatures, bigger than even the largest Alaskan grizzly, were a dangerous threat to the band. They would quite often attack the hunting parties, and to kill a bear was a special event. But this particular bear had left its hibernation quarters long since, and there was no danger that it would return before the autumn.

Ursula's birth was uncomplicated and attended by her mother's elder sister, who sliced the umbilical cord with a sharp flint blade and tied it off. Like all human babies before and since, Ursula announced her arrival with a loud cry as the air was sucked into her lungs for the first time. Within seconds the fresh oxygen was absorbed into her bloodstream and surged round to her brain and muscles to take over where the placental supply left off. Almost immediately she was suckling urgently at her mother's breast, drinking in the natural goodness of the milk. In this milk were also the anti-bodies she would need to fight off infections while her own immune system built itself up. If, as sometimes happened in the clan, the birth had gone badly and the mother had died, this also meant death for the child, for there was not yet any animal milk that could be substituted to sustain it.

Ursula spent only a few days in the cave before it was time for her mother once more to contribute to the main occupation of the clan – finding enough food to live on. The spring camp had been sited carefully, with a commanding view over the wooded slopes below and close to the gorge through which the bison must pass on their way to the summer pastures on the hills. This spot had been noticed only a few seasons before by a hunting party exploring the region from their main base further east. It was already occupied, not by members of another band but by a small group of a completely different kind of human, Neanderthals. The hunting party had given them a wide berth. These were very strong creatures, stocky and built to withstand the cold; but they showed no particular aggression to the new-comers.

When they returned the following year, the camp was abandoned. It was as if the Neanderthals, even though they would have been a match for the hunting party if it came to a straight fight, sensed the power of the new arrivals and feared them, preferring to leave a prize camp and retreat to higher ground rather than risk a confrontation. There were many stories of the Neanderthals in the band's collective mythology, stories that were repeated around the camp fires in winter. They were rarely encountered nowadays but they must once have been more common. In virtually all of the old, abandoned caves, the band came across the heavy hand axes which were the Neanderthals' principal tool. By the standards of Ursula's fellows, these tools were crude and unsophisticated; they worked the same stone as the Neanderthals, but could make much better use of it. For instance, they would strike off thin slivers of flint and sharpen any dull edges by chipping away at them. All the men had to learn how to make their own flint knives and scrapers, but inevitably some were better at it than others – either better at selecting the right piece of flint in the first place, or better at judging exactly where to strike it to create the best flakes. The Neanderthals, from the evidence of stones left behind in the caves, had never got the hang of doing this.

They were strange creatures, whom the band preferred to avoid and who preferred to avoid them. They could certainly hunt, for the evidence was all around. Horse and bison bones littered their old caves, and in one spot, further north, there was a ravine full of the bones of wild animals that looked as though they had been deliberately stampeded off the cliff edge, then butchered where they fell. Occasionally, hunting parties still came across a small group of Neanderthals in the forests or on the remoter slopes. They were very shy and would melt into the trees rather than confront the hunters. For their part, the hunters never attacked the Neanderthals. A few were tempted to hunt them down for food, but there was a great aversion, almost a taboo, to hunting something that was so nearly human.

By the time Ursula was born, Neanderthals were a rare sight. Her ancestors had moved very slowly, over the generations, from the Near East through Turkey. They had crossed the Bosphorus which separated the enormous freshwater lake to the north, now the Black Sea, from the Aegean to the south. In the past, whenever the climate cycles turned and it became colder, there would be a slow retreat towards the Middle East and the Neanderthals would reclaim their lost territory. But this time Ursula and her band had penetrated much further into Europe than any of her kind before her; and unlike her distant ancestors, this time they did not retreat when it turned colder.

Ursula and her band certainly looked quite different from the Neanderthals. They were only slightly taller, but with a much slimmer build, betraying their adaptation to the warmer climates of the Middle East and Africa, where the ability to disperse heat rather than conserve it was the overriding requirement. More than a quarter of a million years of adaptation to the colder European climate had evolved the Neanderthal body shape to a stout and compact form so as to reduce surface area and heat loss. Their faces looked different, too, with a receding forehead, no chin to speak of and bony ridges just above the eyebrows. Whereas Ursula's band had small and inconspicuous noses, the Neanderthal nose was distinctly large and protruding, so as to warm up the cold air before it reached the lungs.

These physical characteristics were not in themselves enough to explain why the Neanderthals slowly began to withdraw as Ursula's band and other modern humans began their slow infiltration of the European mainland. The gradual Neanderthal extinction would take another fifteen thousand years of retreat until the last one died in southern Spain. There were no pitched battles, no deliberate suppression of the 'first nation' Neanderthals to rival the European colonizations of recent centuries. For one thing, the level of political organization required to achieve this was entirely lacking in Ursula's people. They were not a state, with territorial ambitions and weaponry at their disposal; they were just bands of people, living on the edge and just trying to survive. Nor was it their skill with the flints that made the difference. It was the higher levels of communication and social organization that made Ursula's people the better survivors.

Ursula spent her first year being carried by her mother on the daily round of food gathering. A lot of this took place in the woods close to the spring camp. Spring itself was a lean time, for there were no fruits on the trees yet; the band relied on the men to kill at least a few deer or even a bison. It was Ursula's job, as soon as she could walk, to help her mother in the woods. There were frogs to collect by the sides of streams, birds' eggs in the bushes, roots and tubers to dig up with a stick or a piece of red deer antler. Autumn was the best season in the woods: there were hazel and beech nuts to gather in, berries hung from the bushes and there were mush-rooms and toadstools on the ground. The band were often on the move from one camp to another as the seasons changed. Summer would be spent up in the mountains hunting hares and deer, autumn in the oak woods and camped by the gorge to ambush the return-ing herds. Then in winter it was down to the plains before moving up once again to the spring camp. This pattern was repeated year after year after year. Some years were good, the game was abundant and more children survived. Others were less so, and children and old people starved to death in the long winters. Life was very, very hard, and survival depended on a strong constitution and a great deal of luck.

Ursula was one of the lucky ones and did survive. Her mother died at the age of twenty-nine when

Ursula was twelve. By then she had lost some of her teeth and her leg had been badly broken in a fall. The wound became infected and she died of blood poisoning six weeks later. Her comparatively early death did not have a great effect on Ursula's life. She was almost fully grown and was immediately adopted by one of her aunts, who appreciated an extra pair of hands to help with the daily tasks, which had become increasingly exhausting with her own two young children in tow. It wasn't long before Ursula's dark good looks and obviously developing body caught the attention of the young men. They would try to show off, racing or fighting each other to gain her attention. One presented her with a necklace of polished bone cut from the antlers of a roe deer and threaded on to some strands of horsehair. Another gave her a beautifully fashioned flint knife, far too elaborately manufactured to be anything but ornamental. Yet another would visit her whenever he had been out hunting and give her first choice of what he had caught that day. In their own way they were vying with each other to impress Ursula as a good provider, a man who could support her and her future children. Forced to choose between her suitors, she decided on the young man who brought her the ornaments – against a strong recommendation to accept the hunter from her aunt, who had grown accustomed to sharing in the prime cuts he brought home.

The following spring, when she was fifteen, Ursula gave birth to her daughter. Just like her mother had done, Ursula nursed the baby and carried it on her back while she foraged in the forest. Four years later, she had another child, another girl. Both grew up strong and healthy, and Ursula lived long enough to see each of them give her a grand-daughter. She died a few years later, at the ripe old age of thirty-seven. As she lost her teeth she became weaker and weaker as she was unable to chew the tough food that was the staple diet between animal kills. As the band set out once again from the hills to their winter camp she knew she could not make the journey and asked to be left to die in the cave where she and her children were born. Her family was reluctant to leave her, but they also knew the band couldn't afford passengers on its long trek to the coast. So they made her as comfortable as they could and wrapped her in a bearskin to keep her warm. With a last kiss, and with eyes full of tears, her two daughters left her and joined the band on its way down the gorge. As Ursula lay in the cave entrance, looking out over the vast plain towards the distant sea, she thought she could just pick out the small dots that were the band. Or perhaps she just imagined it as she fell asleep. In the morning she was gone. Just the skin, torn and red with blood, remained as a witness to her swift and violent end. The bear had returned.

Ursula had no idea, of course, that both her daughters would give rise, through their children and grandchildren, to a continuous maternal line stretching to the present day. She had no idea she was to become the clan mother, the only woman of that time who could make that claim. Every single member of her clan can trace a direct and unbroken line back to

Ursula. Her clan were the first modern humans successfully to colonize Europe. Within a comparatively short space of time they had spread across the whole continent, edging the Neanderthals into extinction. Today about 11 per cent of modern Europeans are the direct maternal descendants of Ursula. They come from all parts of Europe, but the clan is particularly well represented in western Britain and Scandinavia. Cheddar Man is perhaps the most celebrated of its former members.

16

XENIA

Twenty thousand years had elapsed since Ursula's death. It was now twenty-five thousand years before the present and the world was even colder. The Neanderthals were gone, and modern humans had Europe to themselves. The great plains which stretched from lowland Britain in the west to Kazakhstan in the east were bare of trees save for a few patches of birch and willow scrub on their southern margins. This was a bleak and windy place, with vicious blasts from the expanded polar ice caps sending the winter temperatures down to twenty degrees below zero for days or weeks at a time. Cold and inhospitable it may have been; but the European tundra was also teeming with life and good things to eat. Massive herds of bison and reindeer moved slowly over the plains, feeding on the rich growth of grass and mosses. Smaller herds of wild horse and wild ass were also there to be hunted. But the dominant animal, with no enemies to fear, was the gigantic woolly mammoth. No natural enemies, that is, until the humans arrived.

Xenia was born in the wind and snow of late spring.

Even though it was already April, the snow that covered the land in winter was still on all but the lowest ground and lay in a thick and filthy slush around the camp site. Xenia herself was born in a round hut, about three metres in diameter, whose frame was constructed almost entirely of mammoth bones. Two gigantic tusks formed the door, which was covered by three layers of bison skins to screen the interior from the cold. The gaps between the bones were filled in with moss and soil, and the roof was made of turfs laid across a lattice of willow twigs. In a small hearth in the centre of the hut the red glow from the fire dimly illuminated the inner walls. There was no wood on the fire; all the trees in the vicinity had been used for firewood months ago. What burned in Xenia's hut, and the first thing she ever smelled, was the sickly, unforgettable stench of scorched bone. The tundra was littered with the bleached skeletons of mammoth and bison. They made a reluctant and distasteful fuel, but suffering that foul aroma was better than freezing to death.

The camp was built on a slight rise within a mile of a large, sluggish river. Generations of bison had passed across this river, on the way to and from their summer feeding grounds. Just as Ursula's spring camp had been close to a migration route, so Xenia's was placed to take advantage of this predictable and dependable source of food. Since Ursula's time there had been some technological advances. Flint-tipped spears had been improved, and their accuracy and range increased with the aid of spear-throwers, short pieces of bone or wood that cupped the butt of the spear at one end and acted like an extension to the throwing arm. Novelties and inventions like these were soon disseminated as separate bands congregated at river crossings, or met up while stalking in the tundra later in the summer.

Every year the bison crossed at the same point, where the river curved away, sending the current digging into a steep earth and gravel bank. The migrating herds had gouged a pathway through the collapsing bank, but it was getting steeper every year, making it harder for them to get out of the river. If rationality had entered into it, they would have looked for another, safer crossing; but the same route had been used for centuries, and was not going to change. This blind obstinacy and refusal to adapt, quite the opposite of human virtues, suited Xenia's band very well. As the animals struggled out of the river, exhausted by the crossing and unsteady on the collapsing soil of the earth bank, the spearmen would find an easy target. To avoid being seen and panicking the herd too soon, they had built a hide from mammoth bones with skins to conceal them.

As well as heading for the same place, the herds always came at the same time each year. The band could sense when their arrival was imminent by the lengthening days and the arrival of the geese from the south. The hunting party set off for the river to take up position behind the barricade. When the bison came, they would come quickly. It was no use waiting till they were already crossing the river. You had to be in position first. The first signal of their approach was a faint low sound to the south-east, blown in by the wind like the continuous rumble of distant thunder. As the sound swelled, the adrenalin started to flow and the hunters checked their spears to see that the slivers of flint were securely hafted to their wooden shafts. The drumming of a thousand hooves grew louder and louder. Then the sound of splashing water announced that the lead animals had entered the shallows on the opposite bank of the river, still out of sight. The hunters waited, crouching below the screen for what seemed like an eternity but was in reality only two or three minutes at most, as the animals swam across the river.

At last the first animals, soaking wet but intent on moving ever onwards, came stumbling up the bank and into view. As they struggled to get a foothold in the unstable earth, the animals pushing up hard from behind only increased their panic; but at last the huge red-brown beasts found their footing and began to stream up the bank only four feet from the crouching hunters. Still they waited, until the crush to escape the river had slowed the herd down. Then, from between the hanging skins on their hide overlooking the path, the hunters launched their spears at point blank range into the sides of the animals. They aimed for the neck and chest. The razor-sharp flint tips sank deep into the bisons' flanks. The wounded animals rolled their great eyes and bellowed in pain. They were hardly ever killed outright; the only hope the hunters had of that was if the flints had sliced an artery or punctured the lungs. As the stricken animals charged out once more on to the tundra, the hunters abandoned their hide and followed. With luck the wounded animals would soon collapse and could be safely despatched with a spear through the heart. If they were less seriously wounded they would travel on for miles and die days later out on the tundra.

As each beast succumbed to loss of blood or lack of oxygen, the hunters crowded in for the kill, striking with their spears deep in and out of the chest until the eyes glazed over, the tongue rolled out and the creature was dead. Working quickly with their flint knives, the hunters skinned and butchered the animals where they lay and carried the meat back to camp, sometimes several miles away. At a time of plenty like this there was no need to use every scrap of meat on the carcass, and they took only the best steaks from the flanks and shoulder as well as the liver, heart and kidneys. The rest they left behind on the tundra, only the flint spear-tip still embedded in the great neck leaving any clue for archaeologists millennia afterwards as to how the beast had met its death.

The meat from the bison kills lasted for several weeks as the final snows melted from the tundra and the days lengthened. Geese, ducks and curlew that had migrated from wintering grounds further south to breed on the tundra began to build up their nests among the coarse grass and moss. For a few weeks life was easy; but before long the band would have to strike north to follow the herds. Moving from one temporary camp to another had always been the way of life for Xenia and her band. The most urgent need was to make sure there was enough food over the summer for the band's members to build up enough fat to last through the lean winter months. Xenia's band relied completely on the migrating herds and followed them throughout the summer. There was no wheeled transport, not even sleds, so everything had to be carried. The mammoth bone frames could be left where they were and used again next year, but the skin coverings never lasted more than one winter. There was very little spare capacity, and anyone who was unable to walk on these long marches – the sick, the old, the weak – was left to die. Only when children were old enough to keep up with the band and no longer had to be carried would their mothers conceive again.

Xenia, a precocious girl, had inherited the fair hair and blue-grey eyes of her father. She ran with the other children in the band, helping her mother to organize the camp. Just occasionally she was allowed to join her father in the summer as he went out alone to hunt wild ass. On the rare occasions when he was successful, she helped him to skin and joint the meat. From time to time on these enjoyable forays they would meet up with people from other bands who patrolled the adjoining territories. These were usually friendly encounters, and members of different bands came to recognize and remember one another from previous meetings. They would exchange news, mainly about the weather and the hunting, but also about their families. Their language was not elaborate, but quite sufficiently developed to impart this basic information. Sometimes a young man would go back to another's camp and even stay there for a season. In these small ways, information and people ebbed and flowed across the vastness of the frozen wilderness.

In time Xenia became pregnant. It was a difficult pregnancy and towards the end she could barely move. Though she was a strong girl, even she could scarcely walk as the bulge in her abdomen got bigger and bigger. First her mother and then the other women in the band began to be concerned. Fortunately, they were in their summer camp, the game had been plentiful and they would not need to move again for several weeks. It was not shifting camp that worried the women, but the fact that Xenia was about to give birth to not just one child but two. This was a terrible thing to happen. A mother could never nurse and carry two children at once. That was the whole point of delayed conception, so that until the first child was fully weaned a mother could not conceive another. The hormonal adaptation just simply would not allow it, precisely to prevent this eventuality. And yet, every hundred or so births, a mother produced twins, just as Xenia was about to do. It had happened before, and there was a strict rule in the band that the smaller of the two twins must be killed at once. The only exception was in the rare event that another woman in the band had lost her own child but was still producing milk. But the other babies born that year had all survived.

Xenia herself was unaware of this cruel but necessary tradition, or even that anyone ever had more than one baby at once, for the smaller twin was always killed straight after the birth and the body concealed and buried. But, although Xenia did not realize she was about to have twins, her mother was convinced of it. Unusually, she confided her fear to Xenia's father – unusually, because all matters of childbirth and rearing had always been the unspoken monopoly of the women. He had not known the rule about twins, but agreed with it when it was explained; he was also extremely concerned that Xenia might not survive the birth. Again very unusually, he mentioned the problem to a hunter from another band that he met on the tundra, and who he knew from last season had a daughter about the same age as Xenia. This girl, it turned out, had just given birth to her first child a few days before, but the boy was small and sickly, and was not expected to survive. That evening Xenia's parents hatched a plan. If they could smuggle out one of the twins and give it to his friend, he might agree to take it to his own daughter if she had by then lost her own baby. It was a great risk, because there was no opportunity to get agreement to this in advance.

Later that night Xenia's twin daughters were born. She held them both briefly to her breast before her mother made a swift decision and took one of them outside. She wrapped it in a soft rabbit skin and gave it to Xenia's father who was waiting. He set out at once for the neighbouring camp, nearly twenty miles to the east. It was early morning before he reached it and his friend greeted him. Yes, his daughter's baby boy had died two days ago. Xenia's father held the baby out to him as he considered the proposal. If he did not accept, then Xenia's father would have no choice but to kill the baby. After a few moments weighing up the distress his own daughter felt at the loss of her baby son against the possibility that she might refuse to accept another woman's child, he agreed and carried the now starving bundle to his daughter.

Xenia never knew what happened to her second twin. Nor did she ever know she was a clan mother. The daughter she kept with her started a long line that carries on to the present day in Europe, with about 6 per cent of today's population tracing their maternal ancestry back to Xenia through that branch. The identical twin that was adopted also flourished. Her band and their descendants moved further east in successive generations across the endless steppes of central Asia and Siberia, and eventually joined in the migration into the Americas. Today about 1 per cent of native Americans are the direct maternal descendants of Xenia. Within Europe, three branches fan out over the continent. One is still largely confined to eastern Europe, while the other two have spread further to the west into central Europe and as far as France and Britain.

17

HELENA

Helena lived twenty thousand years ago at a time when the last Ice Age was at its most severe. Glaciers and permanent ice fields covered all of Scandinavia and stretched as far south as the present-day cities of Berlin and Warsaw. The Baltic Sea was permanently frozen, as was the North Sea from Denmark to the Humber. In the winter the Atlantic froze and there was pack ice as far south as Bordeaux. Britain, still joined to continental Europe by dry land, was buried under ice down to what are now the English midlands, central Wales and southern Ireland. Year by year the tundra, the bleak terrain which was nothing more than a thin layer of soil and vegetation above the permafrost, advanced further and further south, almost reaching the Mediterranean. Freezing temperatures and heavy snowfall made the tundra uninhabitable in winter, and the hunting bands who roamed across most of northern Europe were progressively pushed up against the mountains of the Pyrenees and the Alps. Many had been funnelled down the wide valley of the River Rhône and spread out along the low-lying lands that bordered the Mediterranean. As now, lagoons indented the coast, but the shoreline itself was many miles away from its present position. So much water was now locked up in the great ice sheets that the sea level was over a hundred metres lower than it is now.

There was a reasonable living to be made from the shoreline and from the woodland behind. Helena spent her childhood in this landscape, helping her mother comb the woods for wild mushrooms and toadstools, or wading into the brackish lagoons in search of oysters. Her father patrolled the woods alone, on the lookout for small deer and other mammals. But as the first mists of late summer began to hang in the morning air above the marshes, the band knew it was time to prepare for the great gathering.

They packed up their camp and moved inland towards the hills. They travelled light, taking with them only the absolute essentials. Every few days they would come across other bands moving in the same direction. There was no friction between them; indeed, there was instead a shared mood of excitement and anticipation in the air as they moved across the landscape. The woods had thinned out now and they were out into the tundra. They carried on over open rolling hills and flat plateaux and across wide river valleys. At last, after six weeks, they reached their destination, the valley of the Dordogne. The great river flowed green and smooth between high cliffs of yellow grey limestone.

The band were to make their camp in a broad rock shelter that led into a deeper cave. Before taking it over, the men went back into the cave as far as they could go to make sure they were the only occupants. This was always a dangerous and frightening operation. The caves were also used by hyenas, lions and gigantic cave bears. If it was occupied then the residents would have to be either evicted or killed. But this year they were lucky; the cave was empty when they arrived. The camp was set up close to the entrance. The long journey was over. Helena and her companions could rest, warmed by the sun and gazing down at the river as it flowed gently past, a hundred feet below. It was a beautiful sight. Within a few days, all the surrounding caves and rock shelters were occupied as bands from far and wide converged on this magical place. They had come, just as their ancestors had before them, to intercept the reindeer as they made their way from the summer pastures high up in the Massif Central to their winter feeding grounds on the flat plains below. They had to cross the Dordogne and they had to get through the gorge. And Helena's band would be waiting for them.

But this great communal event was still some weeks away, and there was a lot of preparation to do. Helena's father began to strike a new set of flakes from the core of fine-grained flint that he had acquired through a trade earlier in the year. It was very high quality, of even texture with no cracks or other faults. He was an especially good toolmaker and could make almost anything from this precious core. It all depended what was required. He decided this year to renew the bone points on his favourite spear, which would be his main weapon when the time came to kill the reindeer, and settled down near the cave entrance to begin his work. The core itself was a rough cylinder about the size of a small cheese. He looked closely at it, turning it round and round in his hands, calculating by an intuition born of years of experience where best to strike to split off a blade from its edge. It was almost as if he could sense the internal structure of this precious piece of stone, the weakest plane of its molecular bonds. He chose his spot. Taking the core in his left hand and a large river pebble in his right, he struck hard. The rock split, and a long thin blade snapped off from the side, precisely as he had anticipated. While he was in the mood and things were going well, he struck off another five blades before putting the precious core back into its pouch. The blades, about three inches long by an inch across, were wonderfully adaptable. With further delicate retouching they became knives, scrapers and spear points, or tools to work secondary materials like bone or antler. Inspecting each blade in turn, he chose three to use as spear points, one as a scraper for cleaning reindeer hides and two as tools for working bone. Even though any of the six blanks could be shaped for any of the final uses, he knew from experience which blank to select for which end-product.

Today he was going to make a new set of bone points for his spears, and also make Helena's mother some new sewing needles. He still had last year's spear points, but always preferred to make a new set for the coming hunt if there was time. He selected a piece of reindeer antler about six inches long and reasonably straight. These were easy enough to come by in the early summer when the reindeer shed their antlers and began to grow new ones. It meant a week-long trip to the hills behind their summer camp to a place he knew where there were usually some lying around. He could easily have kept some back from last year's autumn hunt, and sometimes he did, but the antler trip to the hills in the early summer was something he always looked forward to. It was a family tradition. His father had taken him every year since he was seven years old, and he had done the same with Helena's older brother. Because of these trips he always had a good supply of antler blanks. He broke off the points and left most of them where he found them, taking back only the pieces he could use, along with a few extras to work up and trade. For instance, he had a deal with a man in the band that he would exchange worked antler goods, which he enjoyed making and for which he had a good reputation, for blade cores. The best flint for the cores came from a long way away, so it made sense that while he collected antler and made it into useful items, someone else trekked off in another direction and collected the flints. So he was quite content, sitting comfortably at the cave entrance, looking down at the river and across to the hills on the opposite bank. Helena, who was now eight years old, came over to sit with him and help out. She had inherited her father's dexterity and was always begging to be allowed to make something.

The first task was to make the burin, which would be used to make parallel cuts into the antler; this required a flat edge like a carpet knife. Helena's father picked up each of the blades in turn and examined them closely. He selected one and laid it down carefully so that one end rested on the ground and the other was lying across a piece of antler. He made careful adjustments until the blade was touching the antler at just the position he wanted it to fracture. Then, in a swift movement he hit it sharply with a small pebble and the end of the blade flew off. It was a perfect fracture and yielded a perfect burin: a good straight edge like a chisel and very sharp. It didn't always work first time, but this one was a real peach. He picked up an antler blank and scored a straight line along its length with the burin. This was a good tool, as good as any he had made. Rotating the cylinder of antler in his hand, he repeated the process until it was divided by the deeply scored lines into five equal segments. It was always hard to get this right, but this burin had cut such a good line that the segments were of exactly equal size. There would be no wastage here.

Slowly he cut down along each of the grooves into the hard bone core of the antler, keeping the lines absolutely straight as he went. This took the best part of an hour. Finally, when he had almost reached the middle of the antler, he pushed down hard with the burin and twisted it. The bone bent slightly and then snapped cleanly along its whole length. Carefully, he lifted out the segment, six inches long and an inch across but now almost triangular in section. This was going to make a good spear point when it was worked up. One by one he split off the other segments. He only had one disaster, when the third segment snapped halfway up: that would do as a needle blank. He gave it to Helena, along with the burin. She already helped her mother to stitch, so it made sense that she should help making the needle. Carefully and evenly she cut away at the splintered segment, smoothing it on each side and tapering it to a point. When she had finished she showed it to her father. It was an excellent first attempt. He took out the awl. This was another of the tools he had fashioned from the multi-purpose blanks, and had a sharp spike of stone protruding from one end. Good awls were extremely difficult to make, and this one was carefully wrapped in its own piece of skin. With the point of the awl Helena's father gouged out an eye at the blunt end of the needle and gave it back to Helena, who ran back to show her mother what she had made.

Good, warm clothing was a must. The winter temperature could stay at minus ten for weeks at a time. Fortunately, there was no shortage of skins and everyone had a made-to-measure tailored outfit. These were layered with an inner skin made from hare, squirrel or anything soft. It was the women's task to make the clothing, and Helena's mother had strong fingers and good eyesight. She trimmed each pelt and matched the pieces before using her own awl to make holes along the edges. Then she threaded the needle with a length of reindeer sinew and, pushing it carefully through each prepared hole, stitched the pelts closely together. Today she was making an outfit for Helena. Children of her age grew so fast, it was hard work keeping up with them. There were no clothes to be handed down from her older brother; he was seven years older than Helena, and they weren't going to carry his old clothes around for seven years. Occasionally she would get a cast-off garment from one of the other women in the band whose child had grown out of it, but on the whole she preferred to make a new outfit from scratch. These clothes had to fit well to keep out the bitter cold, and Helena stood in front of her mother while she was measured up using a long strip of deerskin. The process of joining the pelts, the fittings and stitching the seams took the best part of three days. A well-stitched outfit was something to be proud of, and Helena's mother was keen for her handiwork to be admired. With her prowess as a seamstress and Helena's father's reputation as a craftsman when it came to making antler goods, the family were very conscious of their standing in the band.

By the time they had been in the cave for ten days, they had caught up with the season's tasks. Helena had new clothes, her mother had a dozen new bone needles and her father had a new set of spear points. Already the days were growing shorter and colder; the birch leaves were turning yellow and the first night frosts had dusted the tips of the rushes in the valley below. The reindeer would soon be here. But before they appeared, and to make sure that they did arrive, there was an important ceremony to go through. On the night of the full moon after the first frosts, the men of the band and all the other hunters who had converged on that part of the river made their way up a side valley to a narrow opening in the cliff blocked by a circular stone. Their faces were daubed with red ochre, their bodies blackened by charcoal from the fire. They rolled the stone aside and filed silently into the cave, holding small candles made from animal fat to light the way. Helena's brother was there for the first time. He was old enough to be allowed to join in the hunt, so he must also come to the cave. He was afraid of the dark, and he hated even more being confined in a narrow space. In complete silence the men walked deeper and deeper into the heart of the cliffside, their lights flickering and casting eerie shadows on to the walls. At last, after a good half a mile, the narrow passageway began to broaden and soon opened out into a high cavern. It was absolutely silent except for the drip, drip, drip of water percolating through from above. In places the walls were covered in ribbons of pale flowstone which glistened in the candlelight. On one side three great stalactites, two metres long, hung down from the ceiling while three stubby stalagmites growing from the floor climbed up to touch them, reaching for an embrace that would not take place for another five thousand years.

But these natural wonders were not what the men had come to see. They turned to the right and climbed up into a tall passage leading off from the main chamber. High above them on the wall, barely visible in the pale yellow light of the flickering candles, were the unmistakable forms of wild animals. Huge likenesses of bison, wild horse, reindeer and fierce bulls covered the smooth walls. Helena's brother, already tense with claustrophobia, shrank back and held on tight to his father; he dropped his candle on the floor, it fell into a small pool of water and the flame sizzled and died. A wild bull seemed to be charging right at him, nostrils flaring, head lowered, horns ready to skewer him to the cave wall. Though his father had told him about the paintings, he was not prepared for this. They were so real, so alive and so dangerous. He wanted to run out of the cave, but his father held him tight and stroked his hair to calm him.

In silence the men looked up at these creatures that they not only feared but also depended on. In the soft light the images started to come to life. They began to move. Helena's father rubbed his eyes. Though he had been coming here for twenty years, first with his father, then alone, he always experienced the same strange effect. The paintings were too high up to touch to see whether they had really moved. Still the men stared up in silence, their eyes darting from one animal to another as if to check it was still there. They were concentrating on the hunt, looking hard at these images and preparing to meet them in real life. Nobody knew who had painted these pictures, or how long they had been here. The image of a hand, its outline airbrushed in soot, might have been the artist's – but no-one really knew. Perhaps they had always been here.

After what seemed like a lifetime to Helena's brother, the mood changed. By now the flickering images had become completely real in the minds of the hunters. One by one, they took out their long spears and with a loud cry thrust them back and forward into the air, into the flanks of their imaginary quarry. They did not strike them. They did not need to. The beasts had lifted off the walls and were in front of them, suspended in the air. The cavern echoed to the blood-curdling screams of the hunters as they invoked a sympathetic magic that would transfer the ritual killing of these imaginary beasts to the slaughter of their living cousins.

After a few minutes the noise died down; the men lowered their spears and once again stared silently at their intended quarry. There was no more they could do to make sure that the animals would come and that they would be successful in the hunt. On no particular signal, but sensing it was time, the men filed silently back to the cave entrance and out into the fresh night air. The chill brought them round and they began to talk to each other about the coming hunt. Strategies were discussed, alliances were formed. Helena's brother was just glad to get back out into the open.

Within a week the first reindeer were spotted further up the valley slowly making their way along the gorge. So far so good. They were on the right-hand side, so had to cross the river before they could get through the gorge itself. Helena's band had selected a spot where the river was flowing across large pebbles. It was about fifty metres across at this point with a rocky island in the middle. This is where they would be stationed, using the rocks for cover and hoping that the deer would choose to make the crossing here as they had done in previous years. There were plenty of places upstream, but the presence of the island and the chance it offered to divide the crossing, coupled with the growing urgency the deer felt to get over to the other side before the river reached the base of the cliffs, made this a better place than most. It was only a hundred metres or so upstream from the cave where Helena and her mother watched as the men took up their positions.

This year Helena's father was going to try a spear-thrower and detachable point for the first time. They had been around for a long time, but he had always preferred the traditional design of a hefty wooden spear tipped with a bone point. The advantages of the spear-thrower, as his friends never tired of telling him, were greater distance, greater accuracy and – best of all – that you only lost the spear point and not a complete spear if the animal ran off. The spear-thrower itself was a stout piece of wood, which fitted loosely on to the bottom of the spear shaft and acted as a lever. By putting this over your shoulder and bringing it forward quickly, the spear point itself flew off the thrower much faster than the conventional one-piece spear. The point itself was a sharp piece of bone or flint hafted to a short piece of wood. Since it was also weighted with a stone, the impact when it hit the target carried as much force as a full-length wooden spear. Helena's father had practised with it, but he still wasn't im-pressed. He was really only taking it along on this trip to keep his friends quiet. He was tired of being called a reactionary, so he was going to try it, but he made sure he took his 'proper' spear with him as well.

Helena watched as her father and brother crouched down behind the rocks on the island in the middle of the river. Suddenly a small group of reindeer appeared three hundred metres upstream on her side of the river. They were obviously nervous, sniffing the air and moving their heads from side to side as they walked slowly along the bank. She lay down flat and peered over the edge of the cliff. If the deer saw her they would panic and run back upstream. They moved slowly past the island. Had they sensed the hunters crouched behind the rocks? They came right up to the point, directly below Helena, where the river cut into the cliff. She peeped over the edge and looked down on them from above. She could see their grey backs and their great spreading antlers. She counted twelve of them. She thought they were probably mostly mothers with their calves, but since both male and female reindeer had antlers it was impossible to be sure. There was no way forward for them on this side of the river because the sheer face of the cliff rose directly out of the water. The current had quickened and the water was fast and deep. They waited for a few minutes, unsure whether to chance their luck; eventually they decided against it, turned and headed slowly back upstream. They reached the point opposite the island where the hunters had positioned themselves. Would they cross here or head further upstream? Helena could see them hesitate; then, at last, one of them plunged into the water and began to swim for the island. The others followed. The hunters tensed. Their hearts pounded and their mouths went dry.

As the first of the deer reached the island, the hunters flew at them. At short range the weighted spear points were lethal and accurate. Two deer fell where they stood, blood pouring from their necks. The others charged straight ahead. Helena's father managed to drive his spear deep into the flank of a young calf, then followed it into the shallows and finished it off by cutting its throat with his knife. After the first volley from the spear-throwers some injured deer had turned back the way they had come. The men and boys ran into the water after them and tried to drag them down. Helena's brother foolishly held on to a large adult that had only been slightly hurt. It turned and lashed out with its antlers, catching him with a vicious blow to the side of the head which knocked him unconscious into the river. Watching from above, Helena saw this happen and stood up waving and yelling at her father to attract his attention. He looked up and, realizing something was wrong, scanned the river for his son. By now he was drifting face down towards the fast currents beneath the cliff. His father let go of the calf he had killed and plunged into the river. He reached his son before it was too late and dragged him to the bank, forgetting all about the deer hunt. The boy was soon revived; but the deer were long gone. The dead calf drifted downstream towards the rapids. Nobody was going to get to it in time.

As Helena stared down at it, the river was no longer clear and green but running red with the blood of the slaughter. Judging from the colour of the water coming from upstream, other bands had had a good day; but for Helena's band it had been a disaster. They had managed to kill only three deer, two calves and one adult. That meant a lean winter ahead unless more reindeer arrived. But no more came that way. Two weeks later, the band could wait no longer. The snow had begun to fall, and the other bands were leaving for their winter camps. They packed up and headed back on their long journey back to the sea. If they survived the winter they would be back again next autumn, hoping for better luck.

The years rolled by, following the same pattern dictated by the seasons. Helena's brother was killed three years later, trampled to death by a small herd of wild horses he and his young friends were foolishly trying to ambush. Helena's father lived for another ten years, long enough to see Helena give birth to the first of her three girls. Her mother developed bad arthritis in her fingers, which put an end to her dressmaking, and she died a year later when it spread to her knees and ankles. Helena herself lived until she was forty-two, a very old age in those days and old enough to see her first grandchildren.

Over successive generations the clan that began with Helena became easily the most successful in Europe, reaching every part of the continent. The reference sequence with which all mitochondrial mutations are compared is Helena's sequence. Forty-seven per cent of modern Europeans are members of her clan. We do not know whether this remarkable success is because her mitochondrial DNA possesses some special quality that gives its holders a biological advantage, or whether it is just chance that makes so many Europeans trace their direct maternal ancestry back to Helena and the freezing winters of the last Ice Age.

18

VELDA

Three thousand years after Helena lived and died, the Great Ice Age had tightened its grip still further. Seventeen thousand years ago the plains of northern Europe were completely deserted; all life, animal and human, was compressed into the Ukraine, southern France, Italy and the Iberian peninsula. Velda, the fourth of the seven daughters, lived in northern Spain in the mountains of Cantabria, a few miles behind what is now the port of Santander. The ocean floor falls away steeply here, so the line of the ancient coast is not very different from today, even though the sea level was over a hundred metres lower than it is now. Like so many others before and since, Velda's family depended on the herds of bison and other animals which spent the summer on the high plateaux to the south, but they also hunted in the thick forests which covered the coastal plain. Being positioned between the two resources meant that Velda and her band could maintain a permanent base in and around the area. There was a lot of competition for the best sites, and that gave Velda and her companions an incentive to

288

maintain a year-round occupation. Had they abandoned it for a seasonal migration to the coast or inland to follow the bison, the chances were that they would return to find it occupied by another band. This was not only inconvenient, it was also potentially dangerous. People had been killed more than once in the past trying either to defend or to reclaim a prime cave site.

With most of the caves occupied all year, it was much easier to establish a convincing residence claim; so forced evictions, though they did happen, were largely a thing of the past. However, this did mean that the men were away from the camp in hunting parties for long periods. Velda's man was a good hunter, and even during those times when game was scarce he would always return with something for her and their three daughters. While he was away she would search for food in the woods close to the camp. Her mother, an old woman of thirty-seven, looked after the children when they were too small to come with her. It was hard work tramping the same territory day after day. She knew it like the back of her hand. She knew which streams held small fish, which ponds were favoured by frogs and toads, and where the oak trees with the best acorns were to be found.

Velda was a striking woman, taller than most at five feet four inches, with sparkling deep brown eyes and dark flowing hair which brushed her shoulders as she walked. Her skin was a soft pale brown in winter, but she tanned easily and in the summer her face turned a deep ebony. It may have been cold but the sun shone just as brightly then as it does now. Though a lot of her time was taken up with gathering food, it was not all work and there were bright summer days when she would find a sheltered spot and just lie in the sun for a few hours and reflect on her life. She was close to the other women of her age in the band, most of whom were related to her one way or another, and they spent time together talking about their lives. She was content, even if bringing up three children was a struggle without a man around for a lot of the time. She got a lot of support from the other women and she supported them. Her mother and her elder sister had helped at the birth of all three of her children, just as she had helped her sister and other friends. The men had nothing to do with the births. They were often not around when their own children were born, and it would have been inconceivable for a man to be present at the birth of his child. Thus it was that the women of the band maintained complete control of the process and the mystery of birth. In their hands, they held the future of the band. In exchange, the men supported them by providing food and protection against the wild animals that were a constant threat. Velda's husband was kind and attentive when he was in camp, and it was always good to see him safely back from the hunt, especially if he returned loaded down with meat for the larder. On longer trips he might be away for two or three weeks at a time, depending on how successful he had been. When he had caught as much as he could carry, he came home.

During these weeks when he was away, particularly if all the men in the band had gone hunting together,

Velda felt distinctly vulnerable. The greatest fear was a nocturnal raid by a leopard. She knew of several instances where children had been snatched where they slept. As darkness approached she lit a fire at the cave entrance and withdrew with her children into a natural crevice just to one side, where she settled them on to their beds of soft skins. Her mother had come to live there too, which theoretically provided additional security — though her mother's nerves were not what they were and she snored loudly. Velda slept lightly, waking up every hour or so and making sure the fire was still alight. Only when her man was home could she share the watch and get a good night's sleep.

On some nights she would become aware of animals moving outside in the darkness. It wasn't that she heard them, for they moved without noise; it was more that she sensed their presence. Once she saw two green eyes shining in the pitch-black night only a few feet away as they reflected back the light of the fire. She tensed and clutched the spear she always kept close by, then tossed another branch into the flames. As the sparks flew up, the eyes disappeared as the animal turned its head away. She was counting on the leopard not knowing how few people were in the cave and calculating that an attack was not worth the risk.

Children were rarely killed in a straightforward direct attack. They usually disappeared when careless-ness or exhaustion had allowed the fire to go out. It was often done so swiftly and so silently that no-one was aware that anything had happened until the next morning. That was the worst kind of disappearance, because you did not know for sure whether the child had been taken or had just wandered out of the cave. This had happened to one of Velda's cousins, who had spent days searching for her only child. Was it still alive somewhere out there in the woods? The answer, of course, was no. The leopard had grasped the sleeping girl by the throat, its jaws clamping down with an irresistible force on her windpipe. She could neither breathe nor cry out as the big cat turned and walked effortlessly and silently out of the cave with the child hanging from its jaw. The fear of the night was very real.

Velda and the other women did what they could to console her cousin, but she never really recovered from losing her only child in this terrible way. She sank into a deep lethargy, refusing to eat, and would sit alone on the hilltop staring down at the dark woods below and calling out for her lost daughter. Other women who had lost a child to a wild animal very often had another one almost immediately, so that the grievous blow was cushioned by the new arrival. But Velda's cousin, tortured by the sense that her girl might still be alive, could not take that route. She became far too weak to get pregnant; her man, eventually despairing of her ever recovering, left the band for good. She took to walking through the woods, crying out softly and looking into every bush and behind every tree. Velda and her friends took her into their caves at night, but still she would not eat properly and could not sleep. One day, as winter drew closer, she did not return from the woods until after dark. She did not need to be warned of the dangers, and her friends insisted she must always come back while there was plenty of light. She followed their instructions for a week and seemed to be getting better. Then, one day, she never returned. They never found her body. They did not know what had happened, but they suspected the truth. The same leopard that killed her child had tracked her as well, and pounced from behind as she threaded her way back through the trees. She did not have any strength left to struggle. Soon she was being fed to the same litter of cubs that had devoured her own child.

Velda had a strong artistic streak. Her grandfather had been one of the men who painted the ceremonial caves and she had even tried to reproduce their wonderful images on the walls of her own cave. Her great wish was to be allowed to do something in one of the big caves which were used only for the ceremonies before the hunt. This was a jealously guarded privilege. Not only did you have to be able to paint, you also had to have a convincingly supernatural gift for magic. Since this was virtually impossible to demonstrate, the aspiring artists tended to exaggerate their eccentric behaviour or claim descent from a long line of magicians. Velda expressed her talents as a delicate craftswoman by carving ornaments from bone and, if she could get it, mammoth ivory. The designs she carved were both symbolic and naturalistic, and she would take weeks or even months to finish a piece, often working late into the night by the light of the fire as her children slept. Her most ambitious project was a highly decorated spear-thrower that she was making out of a piece of juniper wood as a present for her man. This was not meant to be used on the hunt itself, but for the cave ceremonies only. Lately, people had taken to having ceremonial weapons rather than the real thing with them to incite the sympathetic magic. It seemed to be much more appropriate and arguably more effective to wave a special weapon on these occasions. Velda worked on this particular piece for the three months of the summer. She wanted to have it ready for the ceremony in the coming autumn. When her man was away hunting she could work openly, but when he was at home she hid her gift in a crevice at the back of the cave. She wanted to keep it as a surprise.

The finished object was utterly beautiful. Along its whole length Velda had carved a group of three bison. You had to rotate the shaft to see the full picture, and yet the proportions were exactly right. One of the animals had its head turned back, licking its flank with its tongue. She paid particular attention to the heads, carefully etching a series of lines to represent the hair standing out on their backs. Bulging lids encircled the large eyes, and the nostrils flared just as they did in life. Night after night she would add an extra detail until, at last, she was satisfied and hid it away for the day her man returned.

He never did. When his friends came back from the hills, they expected him to be back at camp already. After killing a bison, he had left them early, eager to get home. He had taken the best meat from the carcass and set off on the three-day march back to the cave. His companions had waved him goodbye as he headed off down the valley that would take him home. That was the last time anyone saw him alive. When his friends returned to the camp themselves a few days later and realized he was missing, they set off at once back to the hills to look for him. It was very unlikely that he had got lost, for he knew the land as well as anyone. The weather had been good; it was not too cold, so he would not have frozen to death. Occasionally young men might join up with another band they met on a hunt, but never if they had a woman and children waiting behind at camp. He was not feeling unwell when he left his friends. It was a complete mystery. After four days searching the route he would have taken to get home, checking all the rock shelters that were traditionally used as bivouacs, they found no trace of him. On the fifth day they went higher into the mountains to check a large cavern that was sometimes used by hunting parties who were after ibex. It was very unlikely that he would have taken this diversion, especially since he was returning from a successful hunt, but they climbed up to make sure.

About a hundred metres below the cave entrance they came across his body, or what remained of it. Fur clothing lay in a crumpled heap encasing a disarticulated heap of bones and flesh. All the internal organs — heart, liver, stomach and lungs — were gone. The ribcage, stripped of skin and muscle, was still held together by bloodied ligaments. They turned away. They knew it was him. His face was torn off and his skull crushed, but his broken spear lay close to the body. It was certainly his. About fifty yards away lay another body: not human this time, but a large hyena with another spear buried in its chest. That was how he died. Alone and surrounded by a voracious pack of these repulsive beasts he had struck out at his attackers, mortally wounding one and probably injuring others. But there had been too many of them for one man on his own, and eventually he had been overwhelmed and torn to pieces.

They moved what was left of his body and laid it in one of the crevices of a small outcrop, covering it with stones. His closest friend picked up the point of his broken spear, and the group retreated silently down the mountains. Velda knew the worst from their expressions as soon as they arrived back. She took the spear point and hugged it to her breast, weeping uncontrollably. Soon, behind the dark clouds of grief and despair that came down upon her, the seriousness of her situation began to take shape. Having three children to feed without a man bringing in food was not going to be easy. She could not feed her family by herself, and there was not enough in the forest to get them all through the winter. But losing your man, or your woman, was not uncommon. In her situation the usual pattern was to find another mate very quickly, and a beautiful and accomplished woman like Velda would have no difficulty in finding another man, if not from within the band then from a neighbouring group. But Velda never did this. She stayed within her own band and struggled through the first winter by doubling and redoubling her efforts at collecting and storing berries and nuts from the forest. Her children, even the youngest, were pressed into service. The bison hunt that year produced a rich yield and there was a good run of autumn salmon up the river in the valley below the caves. So there was food to spare, and Velda and her children did not go hungry.

Though the band would have looked after her anyway, she began to repay their kindness by giving them small, carved pieces in return. They were just small tokens, portable pieces of magic: an ivory model bison to take on the hunt; a fish to wear on a necklace when wading in the river pools. Her reputation for exquisite craftsmanship spread, and these pieces were eagerly traded with other bands. Through her efforts her daughters all grew up and found mates. Two left the group and one remained behind, and they shared the cave she grew up in. As she passed into her thirties, older but still striking, she eventually achieved her ambition and was allowed to decorate part of one of the ceremonial caves. She died peacefully in her sleep, at the age of thirty-eight, from a combination of old age and exhaustion. When her daughter found her body, cold and peaceful in the morning, she also found two objects lying beside her on the skins she used as a blanket. One was an old spear point, worn smooth by years of handling. The other was the most beautiful carved juniper wood spear-thrower that anyone had ever seen.

Today about 5 per cent of native Europeans belong to the clan of Velda; they are more frequent in western Europe than in the east. Many of Velda's children have travelled a long way from Velda's home in the hills of Cantabria. A small group found their way as far north as it is possible to travel, reaching the very top of Scandinavia, where they are to be found among present-day Saami of Finland and northern Norway.

19

TARA

Velda and Tara both lived at roughly the same time, seventeen thousand years ago, in the depths of the last Ice Age. They may even have been exact contemporaries; but they certainly never met and their lives were very different. Velda lived in Spain, while Tara's homeland was the hills of Tuscany in north-west Italy. Velda, and Helena before her, were relatively well off. They both lived in a world where the predictable seasonal migrations of the large tundra animals brought fresh meat almost to the doorstep. This abundance led to a relative affluence, and the human population increased. At the annual gatherings of the reindeer hunters there were frequent interactions and exchanges between bands, and a flourishing artistic culture grew up. Beautifully carved ornaments and lucky charms were made from all sorts of raw materials – wood, ivory, shells and bone. Hundreds of caves were decorated with the brilliant and haunting images of wild animals.

Tara's world was much less prosperous although, ironically, it was warmer. The higher temperatures meant that the landscape, other than the highest hills, was heavily wooded. The tundra animals were not there. Instead, the forests were home to red deer and wild boar. These were hard and sometimes dangerous to hunt. Though the woods provided plenty of scope for foraging, the absence of a predictable supply of big game meant that the land could support far fewer people than Velda's Cantabria or Helena's Dordogne. This relative poverty had stifled the growth in artistic expression and the patterns of social exchange. Bands were more self-contained, about twenty strong, and had to work much harder for their food. They were always on the move as they exhausted the meagre harvest of the woods. This was Tara's life.

Her own mother had died when she was ten and her brother only six. They were cared for by their mother's sister, and shared in the daily routine of foraging in the woods. Their father still brought in what he could – a wild piglet, a pine marten, a small roe deer, or, if he was very lucky, a red deer. Killing a red deer was a cause for celebration throughout the camp, and everyone had a share of the meat around the fire. Tara had kept her mother's flute and played a lively tune on these rare but joyful occasions. Her father had made it years ago from the wing-bone of a swan by drilling holes along one edge, one to blow across and three for the fingers to change the notes. The range was limited and the sound rather breathy, but it added to the atmosphere around the camp fire as they sang and danced late into the night. Next day everybody slept late. For once, the daily grind could wait.

As summer passed into autumn, they made their way slowly down to the lower ground, along the valley of the Arno and downstream to the coast. This was twenty miles further than the same journey today because of the lower sea levels. Out of sight, beyond the horizon, the uninhabited islands of Corsica and Sardinia were joined to each other by dry land. Tara enjoyed the sea and walked for miles along the broad sandy beaches, picking up driftwood and anything else that caught her eye that might be of some use. She loved collecting seashells and always brought handfuls back to the camp every night. She would gouge holes in them with a sharp stone and thread them into a long necklace with seaweed or strands of marram grass, knotted together. They didn't last long as ornaments because the thread soon broke, but that wasn't the point – and it gave her a good excuse to go down to the shore again to collect more shells.

One day on her walks along the shore she saw in the distance a large grey shape lying just above the water-line. As she got nearer she could see it was the carcass of a beached dolphin, its jaw open wide showing its sharp and regular teeth. It had definitely not been there the day before, and was quite fresh. Seagulls were already on the scene, pecking at the eyes but making no impression on its thick skin. Even though she had never seen a dolphin before, Tara knew at once that this was food and ran back to tell the others. Everyone stopped what they were doing, gathered up their knives and headed up the beach. The young men, women and children ran as fast as they could, the middle-aged walked, and those over thirty-five stayed behind, remembering what it was like to be young.

As they rounded the headland to the bay where Tara had seen the carcass they stopped in their tracks. There were other humans already there. They had started to cut into the skin. They looked up when they saw Tara's band in the distance and stopped what they were doing. This could turn nasty. There were only five of them – two men, a woman and two children – against ten from Tara's band. If it came to a fight, they would lose. A dolphin carcass was a valuable prize, but not worth dying for. There were strict conventions, universally understood, that a hunter always kept what he killed. Likewise, a carcass belonged to the band that found it. Normally Tara's band would have turned back at this point, acknowledging that they were not the first to arrive. But Tara was the one who had found the dolphin.

Tara did not know the rules, but she sensed that she might be forced to abandon her prize and began running towards the group who were threatening to deprive her of it. Her father shouted at her to stop, but she kept going. Dropping everything except a short spear, he rushed after her. The others followed. The three adults by the carcass stood their ground. Tara had always been a fast runner and her father, fit though he was, was gaining on her only slowly. She was only three hundred yards away from the carcass. Two hundred yards. One hundred. The group by the water raised their spears. Fifty yards. With a final burst of speed, Tara's father grabbed her by the shoulder and brought her down in a bundle on the soft sand. Immediately he was up again and shielding Tara. He faced the spears of the two men who had rushed forward. He was still a long way in front of the support team and in great danger. They were only feet away when he recognized the face of the tall fair one on the left. It was his sister's man. He called out his name. The others stopped in their tracks. An enormous grin spread across the face of the fair one. He dropped his spear and rushed up to Tara's father and embraced him. The relief shone through on everyone's face as the adrenalin ebbed away. The others caught up. Tara spluttered out how she was the one that found the dolphin and pointed by way of proof to a set of footprints in the sand that led back in the direction of her camp. But the men had already agreed to share the spoils. There was enough for everyone, and anyway they had to work fast. The tide was coming in.

Tara's aunt arrived with the other members of her band and the process of stripping the carcass began. Every so often they had to haul the carcass further up the beach as the incoming tide threatened to take it back out to sea. Relays of children carried the butchered meat to a safe place in the dunes above the high-water mark. By the time they had finished, the great orange sun was setting over the sea. It was a still night and they all decided to camp where they were and share a meal on the beach. There was soon enough driftwood to start a fire, and a spit was hastily assembled to rotate the chunks of dark red meat. Their faces illuminated by the soft glow of the fire, the members of both bands renewed acquaintances. Tara was too young to remember her aunt, and her father had not seen his sister for several years since she left the band. Now he sat down with her and told her of the tragic death of Tara's mother two years before and how much he missed her. Why not bring Tara and her brother and join our band for a while? his sister suggested.

That is how Tara moved with her brother from one band to another who hunted further up the coast. Four years later she was pregnant and the first of her two daughters was born. As soon as the baby appeared it was obvious she had inherited her father's flame-red hair. By the time she was a year old it was also obvious that she had inherited Tara's independent streak. She refused to listen to any instructions and was always putting pebbles and even sharp flints into her mouth. Tara was a diligent mother and a welcome new member of the band. She had a good man and the hard life was as enjoyable as it could be.

She looked forward to the winters spent down by the sea. She was always the first to volunteer for beach-combing and, with her daughter on her back, she would walk along the shore for miles, day after day. She knew every rock, every stone, every patch of sand, and spotted at once if the sea had cast up anything new. She liked the wild weather best, with the spray blowing off the waves driven inland by the fierce west wind. After these storms, which could last for days, was the best time for beachcombing. She was out at first light, eager to discover what new treasure the sea had flung on to the land. After one particularly vicious storm, and with the wind and rain still blowing in her face, she came across a long tree-trunk, bleached by its time at sea and thrown up on the highest point of the beach. It had obviously been in the water for a long time, because barnacles had attached themselves to the wood – but only on one side, which seemed odd.

The next day she came back with her father. Even though it was a large trunk, about three metres long and half a metre across, they could move it a little if they both put their backs into it. What made it so light? One side, the side with the barnacles, was hard and polished by the waves. The other side was pockmarked and soft. Tara dug into this with a flint. It came away easily. They carried on scraping out the soft parts, which must originally have been diseased, until they had hollowed out the whole log. This was still heavy but, with a few friends who had joined them, they could carry it quite easily. And, of course, the first thing they did was to launch it into the sea and start throwing stones at it. The water was calm by now and the log floated easily on the smooth surface. But it always floated the same way up, with the opening above the surface and the barnacles underneath. This was very curious, but it did mean there was an added dimension to the game: one point if you hit the log, but two points if you landed a stone inside it.

After a while everyone got bored with this game and began to go home. For no particular reason, Tara and her daughter stayed behind. They were at the edge of the bay where it ended in a low rocky outcrop. The log drifted along the shore until it came to rest against the rocks. Tara and her daughter followed it, sat down and idly threw some more stones at it, many of which landed inside because it was now so close. The log was still floating but there were at least twenty stones inside it. Tara then wondered what would happen if she put a much larger rock into the log. She picked up a big grey stone and carefully placed it in the opening. Surely this would sink it. But it didn't. In fact it seemed to stabilize the log even more.

She had a flash of inspiration. She called her daughter over and lifted her into the log. It settled lower in the water but it still didn't sink. She pulled the log right up to the side of the rocks and stepped in herself. They were floating. She pushed off from the rock and the boat, for that is what it had become, slid slowly across the clear water. She knelt down and paddled instinctively with her hands. The boat slowed and began to change direction. This was fantastic. Over the side she could see the white patches of sand and the dark rocks of the sea bed. She had to be careful not to overbalance the boat and sensed when it was beginning to roll. After twenty minutes she realized she had been carried along by the current into the next bay. With a few movements of her hands she drifted on to the sandy shore and leaped free, pulled the boat up on to dry land and lifted her daughter off.

Fortunately, the weather was still calm the next day, and the boat was still on the beach when she returned with the rest of the band. The children played in it, the men raced in it. Someone produced a flat piece of driftwood and used it as a paddle. At the end of the day

Tara and her man paddled the boat down the shore to the camp and pulled it to safety above the tide line. Other bands came that winter to admire the new plaything. It had no obvious immediate use other than for fun. Only later was it used to reach offshore islands and cruise the shallow waters of the river estuaries on the lookout for flatfish and eels. In the late spring they hauled it high up on to the beach and left it as they headed inland for the summer hunting on the higher ground. That autumn, Tara's second daughter was born: not red-haired like her father, but with her mother's straight dark brown hair. But like them both she had bright blue eyes, quite unusual in the band, whose eyes were more commonly a light hazel brown.

The boat was still there when they returned in the early winter, still seaworthy but a little decayed. The men began to make new ones from freshly fallen timber. It was hard work; most of the trees were either too rotten, which was why they had fallen down, or too tough if they had been blown down in a storm. The next spring Tara, who so loved the sea, suggested to the band that instead of going into the hills, they should stay down by the shore, build some more boats and use them for fishing in the shallow waters and creeks around the coast. Two more families agreed to give it a try, and they spent the whole year moving up and down the coast in the new craft. The men hunted deer and wild pig in the marshes, and the women and children picked limpets and winkles from the rocks at low tide. When the hunting deteriorated in one place, they moved easily along the coast to another. They discovered offshore islands with rocks covered in steel-blue mussels. Seals also visited these islands to haul out or to breed. They made easy prey for the hunters, who could drift up slowly without disturbing them, then clamber ashore and club their victims before they could slip into the water. This maritime life suited Tara. They did not depend on the sea, because they could always head for the woods and the hills; but they were making a living from it, and it made a change from grubbing around on the forest floor. And it felt safer too.

Tara had one more child, a boy. All three were healthy and lived long enough to have their own children. Tara saw her first three grandchildren, all girls, before she died one winter close to the beach where she had found the dolphin all those years ago. She was buried in a grave dug into the sand dunes. Her face was reddened with ochre, as if bringing colour to her cheeks would somehow revive her. Around her neck were placed a dozen strands threaded with hundreds of pierced seashells. She lies there now, twenty miles off the coast of Livorno, under the blue Mediterranean, while a hundred metres above her descendants glide to and fro on their own updated versions of her hollowed-out log.

Today just over 9 per cent of native Europeans are in the clan of Tara, living along the Mediterranean and the western edge of Europe, though they are not restricted to these regions. They are particularly numerous in the west of Britain and in Ireland.

20

KATRINE

Piazza San Marco in Venice is flooded again. The sea
gurgles up through the stone sluices and the super-
intendent wearily orders the wooden duckboards to be
unstacked and laid across the square. Nothing, not even
the sea, must stop the tourists from filing through the
Basilica and the Doge's Palace. Venice is slowly sinking
into the sea. Fifteen thousand years ago, when Katrine
lived there, the sea was over a hundred miles away.
The Adriatic is a very shallow sea, and the worldwide
lowering of sea level towards the end of the last Ice
Age shrunk it to half its present size. Katrine could
have walked in a straight line from Split in Croatia to
Ancona in Italy without getting her feet wet. She lived
in the vast wooded plain that stretched from here to the
Alps and took in the wide Po valley from Bologna to
Milan and Turin. Had it been colder, this would have
been an area of open tundra crowded with wild horse,
bison, reindeer and mammoth. But the relative warmth
of the southerly latitude meant the forest could survive.
The woods themselves were much like Tara's, a larder
of wild food if you knew where to look and were

prepared to put in the work to find it. However, they were much more extensive and the sparse human population was spread out over a much larger area. People still lived in bands, and these bands tended to stick together as they moved through the woods. Katrine's band lived in the northern part of the forest, where it backed up against the steep ramparts of the Alps. Towering above the plain, their snow-capped peaks and vast glaciers, much more extensive than today, appeared to Katrine as a distant and forbidden world.

She had always been a beautiful child, with fair hair and greenish-brown eyes, and she was not far into her teens when she became pregnant by a friend of her older brother. In the summer before the birth the band moved up into the mountains to hunt ibex and chamois. Her mate was still inexperienced in the mountains and not used to the dangers of hunting at high altitude. He was stalking a group of chamois across a cliff, hoping to surprise them and drive them off the sheer face, when he lost his footing and fell four hundred feet to his death. He had always been an impetuous and boastful youth, and the group greeted his death with as much irritation as sadness. Just as he was going to be able to start repaying the group for its years of support by bringing in food, he had got himself killed.

Annoyance was also Katrine's considered reaction. By his foolishness, he had left her with the prospect of bringing up a child alone. She was determined to find a replacement as soon as possible. The baby girl was born in late October, by which time they were down from the mountain and foraging in the woods again. She was a sweet enough child, with her father's dark brown eyes, but Katrine never bonded with her from the start. Just looking at the baby suckling at her breast filled her with intense irritation. Why had she been left with this mewling infant by a feckless man who should have thought of her and the baby before he put himself in danger? But there was nothing to be done. She couldn't palm it off on anyone else. No-one else was lactating and no-one had lost a child.

Her mother realized that there was something badly wrong between Katrine and her baby but couldn't offer any real solution. Until it was fully weaned, which would not be for at least another three years, there was nothing to be done. As the child grew and began to crawl, then walk, matters did not improve. In every new development — the way she smiled, the way she waved her arms — Katrine saw nothing of herself but only reflections of the irresponsible and now loathed father. At long last, after four interminable years, the infant was fully weaned. Katrine had not entirely wasted her time during the long wait. At every opportunity she would leave the child with her mother and seek out the company of her brother's older friends. Over the three years she slept with all of them at one time or another but, because she was still breast-feeding, she never got pregnant. Her mother had realized what was going on for some time and had warned her against such foolishness. Her father didn't seem to care.

So, of course, the inevitable happened. She did get

pregnant again, almost immediately after her baby was weaned. The father could have been any one of the three boys, and she had no idea which one. It was inconceivable that she could have another child without a proper mate, so her mother took her to one side and begged her to identify the father. She refused even to tell her mother who the three candidates were. Her brother was no more forthcoming. It was a desperate situation. Katrine's father, who was not getting any younger, was already having to provide for two more people than he had bargained for; another one would bring yet more responsibility. Though he loved his daughter, he shook her hard to get her to reveal the father's identity. Still she refused. And none of the three came forward when the news of Katrine's pregnancy spread around the camp. No great surprise there.

When the baby was born, Katrine's mother lifted it and gently gave it to Katrine. She looked at it, expecting to feel equally repulsed as she had been the first time. But she was not. As she took the tiny girl into her arms and held it to her breast she was overcome by a feeling of warmth and tenderness. She felt none of the exasperation and irritation that she experienced after the birth of her first child. Though her situation was arguably more desperate now than before, there was no resentment. None of the men had come forward to help her; but here was something utterly helpless, who needed more help than she did. Her attitude to her second daughter was completely different. There was no logical explanation for this transformation, but there was no doubt Katrine had undergone a fundamental change. She nursed the baby carefully and conscientiously. She left it with her mother only so that she could resume her work of collecting food in the forest. She even began to grow closer to her first daughter. Rather than seeing her as a millstone round her neck, a burden and a nuisance, she began to feel much more protective towards her as well. There was no obvious reason for this abrupt change in Katrine, but it had good results. Her father and her brother did not mind the added burden of the extra mouth to feed now that Katrine had resumed her work in the forests. When the next summer came and they once again climbed up into the mountains, Katrine even wished she could join them on the high slopes. This would have been inconceivable a year earlier, when she showed no interest at all in helping anyone but herself. But it was too soon for that. Her baby was still at the breast and needed to be fed every four hours.

While her father and brother were high in the mountains, a very strange thing happened at the base camp in the pine forest below the snow line. It was a dark, moonless night. Katrine and her mother were both sitting close to the fire. Both children were asleep, the elder daughter with her head on her mother's lap, the baby resting beside her on the soft ground. Just as Katrine was about to settle down for the night herself, she thought she saw something move in the forest, about ten metres away on the other side of the fire. The woods were still a dangerous place, with lynx, wolves and bears all active at night. She looked deep into the forest but saw nothing and settled down to sleep.

The next night the same thing happened. She called her mother, but she couldn't see anything either; her eyes were not as good as they used to be. It moved again. There was definitely something there. Katrine strained her eyes and shifted her position to see around the flames. Now she could get a better view if it moved again. But there was still nothing. She moved ten yards away from the fire so her eyes could get used to the dark. After a few minutes she thought she could make out a pale grey shape among the rocks. Then it moved again. Very slightly but definitely. She stared again. There, with its paws outstretched and lying quite still, was a fully grown wolf. She let out a piercing scream. In one swift movement the wolf was gone. Katrine ran back to the safety of the fire. By then everyone was awake, expecting an attack from the dark. Katrine calmed down and then told them what she had seen. It was very unusual to see a wolf so close to a human camp. There were plenty of them around; you could tell that from the howls that echoed through the dark valleys. Occasionally you would sense you were being followed, and turn round to see the long-legged shapes loitering in the distance. They did not retreat, but just stared back, as if to say 'Be careful.' But, in truth, they rarely attacked humans, certainly not humans together in a group, and never near a campfire. Everyone agreed that Katrine must have dozed off and been dreaming.

They changed their minds when next night the wolf was there again, sitting quite still on a patch of grass in front of the same large boulders. It was alone as far as anyone could tell. One of the men walked slowly towards it. It stayed where it was until he got to within twenty yards, then got up and trotted quite calmly further back into the dark. What did this creature want? It was obviously not about to attack them, but what reason could it have for just sitting there and looking at them? The same performance was repeated the following night.

By then Katrine's father and brother had returned from their hunt with a chamois each slung across their shoulders. These were quickly butchered, and before long the spit above the fire held a dozen pieces of venison roasting in the flames. Nobody saw him arrive, but the wolf was back. Katrine's father picked up a piece of raw meat in one hand and, with a spear in the other, he walked slowly towards the animal. It moved its head from side to side as if trying to decide whether or not to flee. Twenty yards from the animal, Katrine's father laid down his spear and crouched on the ground. He moved slowly forward, talking softly as he went, until he was only a matter of twenty feet away. The wolf was getting more and more restless at every step. But still it didn't flee. Gently, and without a sudden movement, Katrine's father tossed the meat to one side of the wolf then, still facing it, moved slowly backwards. When he was almost back to the camp fire, the wolf got up, went over to the meat, sniffed it quickly, then took it in its jaws and trotted off.

They all looked at each other in silent amazement for a few seconds, then burst into spontaneous conversation. One of the men had heard of a similar happening many years ago at a camp in the mountains to the east, but he had never believed it. There seemed no explanation for the wolf's behaviour. Over the next few nights, the animal returned to the same position and took the food that was thrown for it. It started to appear in the daytime too, and would walk behind the hunters as they went off into the hills. As the weeks passed it became more and more tame, coming much closer to the fire and eventually taking meat, gingerly at first, from the hand. Then one night it did not return. The band were disappointed. They had got used to their strange companion. But after a while they forgot about it and carried on with their normal routine.

About six weeks later Katrine's father and brother were returning from another successful hunt when they sensed they were being followed. They turned around and there, standing quite still on the path, was the wolf. Beside it were two cubs. It was not a him after all. The she-wolf and her cubs followed them to the camp and settled down near her old spot. Was this the reason for her visits to the camp? Did she sense that she could be spared the rigours of hunting for her cubs? She certainly accepted food and, when they were old enough to take it, fed them directly from the scraps. Over the next few weeks the wolf was the band's constant companion and her cubs played with the children on the floor of the forest. When the time came to move the camp down to lower ground she did not appear to want to follow them down to the plain, but seemed to want her cubs to go with the humans. She would turn them away and push them back to the camp as it was being dismantled. Katrine understood what she meant. She bent down and picked up the two cubs and carried them away.

During that winter on the plains, the wolf cubs grew fast on the scraps they were thrown. They followed the hunters everywhere and even joined in the chase, bringing down a roe deer or a wild boar that had been injured by a spear. They were certainly earning their keep. The other bands that they came across on the plains could not believe their eyes when they saw the wolves at the camp. So the old stories were true. The wolves stayed with the band that winter, helping to track game and forming an ever closer bond with Katrine and her family. The next summer, when the band went up once more into the mountains, the cubs, now fully grown, became more and more restless and would sometimes leave the camp after dark and not return until the next day. They were torn between their new life with the humans, a safe life that meant a steady supply of food, and the call of the pack whose haunting cries echoed around the valleys. One day they did not return.

Katrine and her band never forgot their encounter with the she-wolf and her cubs. The same strange meetings between wolf and human were replayed many times. Sometimes the cubs would stay with the bands from one year to the next. Little by little they came to depend on the humans and gradually lost their wild instincts as they became the first animals of many to accept a life of domestication. They became dogs. By eight thousand years ago, dogs had become the indispensable companions of the hunters who ranged over Europe after the last Ice Age. Some became so precious that they were given a ceremonial burial with their owners.

Katrine's clan flourished in northern Italy and beyond. Ten thousand years after she lived, one of her many descendants died crossing the Alps. We know him as the Iceman. Today 6 per cent of native Europeans are in the clan of Katrine. As a clan it is still frequent around the Mediterranean but, like the others, draws its present-day members from all over Europe.

21

JASMINE

Compared to the hardships and uncertainties of the lives of the first six women we have encountered, Jasmine had a much easier time. For one thing, she lived in a permanent settlement, one of the first villages. But the accommodation could not be called luxurious by any stretch of the imagination. She lived in a circular hut, dug partly into the soil, with wooden stakes supporting a thatched roof made of reeds. These huts were tiny and cramped; but they were home. The village had a population of about three hundred people, very much larger than any of the temporary hunting camps which were home to the other six women. The village was about a mile from the River Euphrates in what is now Syria. The Euphrates carried the rain and melted snow from the mountains of Anatolia in the north through wide grassy plains to join the River Tigris on its journey to the Persian Gulf.

The Great Ice Age was at an end. The ice caps and glaciers had been melting fast for the past four thousand years as global temperatures climbed erratically toward present-day levels. The water that had

319

been trapped in these great reservoirs of ice now flowed into the ocean basins, so that sea levels were rising around the globe. The low-lying plain that lay between Arabia and Iran was flooded as seawater seeped inland past the Straits of Hormuz to create the Persian Gulf. The Adriatic pushed the shoreline further and further north towards its present position in the lagoon of Venice. Seawater rushed through the Bosphorus and poured into the Black Sea. Britain and Ireland began to lose their connections to the European mainland and to each other as water flowed into what are now the North Sea, the Irish Sea and the English Channel. On the other side of the world Australia and New Guinea, which had been joined together as Sahulland, were separated as the Torres Straits filled with water. The flat plains of Sundaland that once connected Malaysia, Sumatra, Java and Borneo into a single land mass were now seabed. The crucial land bridge that connected Asia and the Americas finally sank beneath the cold waters of the Bering Straits.

All these lands were inhabited, and had to be evacuated as the sea level rose. This was not the gradual process once imagined, with imperceptible advances measured in fractions of a millimetre per year. It now appears that the sea rose in a series of rapid stages, by several metres over only a few decades as water was suddenly released from the melting continental ice caps that had become vast freshwater lakes, their outlets to the sea blocked only by frozen tongues of ice. One such tongue lay across the opening of what is now the Hudson Bay, holding back an enormous inland lake that covered most of Canada. When this ice barrier was finally breached and the water gushed out into the ocean, the sea level rose around the world by half a metre overnight. Sea-level rises of this magnitude today would not only drown millions of square miles of low-lying land but would inundate many of our coastal and estuarine cities. If this version of events is accurate, then the sudden end of the Ice Age brought tragedy to the inhabitants of the coastal plains. Many would have drowned or seen their livelihoods destroyed. Great flood myths permeate many mythologies. Perhaps this is their foundation.

Jasmine's village was safely above the encroaching waters of the Persian Gulf. It had grown up to take advantage of another seasonal migration – not of the bison and reindeer of the tundra, but of the Persian gazelle. The village lay close to the route of their annual spring migration from the hot deserts of Arabia to the grasslands of the hills that encircled this gentle land. The meat they provided could be dried and kept for several months, but would not last out for the whole year.

Jasmine collected acorns and pistachio nuts from the woods nearby, but her main occupation was looking after what she called her experimental plot. For many years now, when the young men followed the gazelle up into the hills, they had kept themselves going by munching the seeds of the wild grass that grew there. Though they needed a lot of chewing, to the young men they had one over-riding advantage: unlike gazelle, they couldn't run away. Jasmine's man was not a good hunter. She had known him as a child and watched, helpless with laughter, as he tried to throw a stone at a pretend gazelle. He was hopeless. The only time he ever hit the target was when he threw the stone underarm. 'Nobody throws spears underarm,' his father would shout. He got a bit better as he got older, but it would be a miracle if he ever got close to killing a gazelle. And he didn't. He never managed to bring down a single one. No-one, certainly not Jasmine, was to know that he had a hereditary weakness in his shoulder which meant he could never improve. But what Jasmine liked about him was his curiosity and intelligence and his kindness. He had a gentle temperament which she found appealing, and although she was concerned that he might not become an extravagant provider for their family – Jasmine wanted lots of children – she somehow believed they would get through.

While she was nursing their first baby, he followed the other men into the hills after the gazelle and wild sheep. He took his spear with him but had no illusions about killing anything; it was just to look the part. His real intention was to collect and bring back to the village as many wild grass seeds as he could. He had taken with him two large sacks made of stitched gazelle skin. He found a hillside where the grass was thick on the ground and the seed heads were already ripe. With one hand he gathered up a bunch of grasses, held them in the mouth of the sack and shook them hard. Most of the seeds fell off the heads and into the bag. It only took him an hour to fill both sacks, and he walked back to the village while his companions were still trying to kill their first gazelle.

When he got home his first job was to try to break off the brittle hairs that were still attached to the seeds. He did this with the grain still in the sack, rolling a large stone round and round on top of it. Then he poured the contents out on to the ground. The hairs blew away in the breeze and left a good pile of largely hairless seeds. He soaked these in water for a few hours, then handed Jasmine a handful. They were hardly delicious, but they were all right – though the husks still stuck in her teeth. He tried grinding the dried seeds between two stones, and this did crack off at least some of the hard outer skins which, like the hairs, dispersed in the wind. But he saved his best piece of ingenuity till last.

He had kept back a few handfuls of seeds to see if he could grow them near to the village. He already knew that the grains germinated into new seedlings. People had been bringing back pouches of wild grain for years, though not on the same scale, and he had noticed how seeds dropped accidentally on a damp patch of ground would soon produce a small green shoot which in time became a new plant with its own seed head. But he was going to try to grow wild grass systematically. With Jasmine at his side he walked down towards the river and found a piece of level land a few hundred yards from the near bank. It had a light covering of weeds and he set them alight to clear the ground. Then he took a stone scraper and scored a line in the soil. He put in a row of seeds and kicked over the topsoil to cover them up; he already knew that the village sparrows had developed a taste for grain. He planted ten rows before he had exhausted his supply of seeds and they headed back to the village.

Next day they returned to the plot. It was exactly as they had left it. It rained for the next few days and still nothing happened. Then, a week later, Jasmine took her baby down to the plot – and there, struggling out of the ground, were ten rows of tiny green shoots. She rushed back to tell her man, but he had not yet returned from another fruitless hunting trip. From that day on, Jasmine and her family spent as much time as possible by the plot. Together they cleared some more land and planted more seeds from the hills. They planted whatever could be eaten. Wild varieties of chickpea and lentil joined the original wild wheat. They showed off their plantings to the rest of the village, who expressed a range of views from the enthusiastic to the downright hostile. They didn't claim that it would replace the gazelle or the pistachio as their staple diet, only that it would supplement it and make them less dependent on one food source. There was no denying that the grain growing on the plot could be eaten. Grinding it between large stones and separating the husks made the resultant mash far more palatable.

Jasmine and her man had also noticed that some of their plants produced seeds that stayed attached to the stem. This was after a fierce wind had stripped the seeds from most of the plants and severely reduced the yield. But a few plants had withstood this battering. On these plants, the seeds were stuck to the stem with less brittle attachments. When these seeds were planted, they wondered, would they grow into similar plants? So they tried it. And it worked. Little by little, year by year, they selected the plants with the attached seeds, the plumper grains, the stouter stems and took their seeds for planting. Within only a few years, the wheat in their plot no longer looked exactly like the wild varieties. It had been artificially selected for the most desirable properties.

By now most of the sceptics in the village had changed their minds, especially after the year when the gazelle failed to appear. A few other enthusiasts had taken to planting out their own plots using seeds given to them by Jasmine. Visitors from nearby villages were equally impressed, and begged Jasmine to let them take a few seeds back with them. The idea quickly spread around the region. By now Jasmine's man had given up pretending to hunt altogether. He was enjoying the sedentary life. They had five children, rather too many for his liking, but what could he do about it? Jasmine just kept getting pregnant. Even before her first child was completely weaned she conceived again. At least there was now sufficient food coming off the plots, which they had enlarged many times since they had started.

They heard that someone from another village, six days away to the north, had found a way of keeping wild goats. Apparently they had captured two kids on a hunt and taken them back as pets for the children. When they grew too big to play with, instead of killing and eating them, which had been their original intention, they tied them to a wooden stick to prevent them escaping and let them browse on whatever vegetation they could reach. A year later one of them produced a kid. Now they had a dozen goats of various ages. When they needed meat, they killed a goat. It was a lot easier than hunting them. The idea of growing your own food was definitely catching on.

Things were going very well for Jasmine and her family. They had a large plot by the river and took on some of the other women and children of the village to help them, rewarding them with a share of the produce. More and more people took up this new way of life. It had great appeal. Anyone could join in – children, mothers with children, grannies. There was always some job to be done, whether it was getting rid of weeds, doing a bit of watering or clearing a new piece of land. You didn't have to depend entirely on the harvest because the oak and pistachio trees were still there. The gazelle could still be hunted. It was a perfect combination.

As Jasmine sat looking at their field with the wheat ready for the harvest, little did she realize that she and others like her had started a revolution that would change the world for ever. Within only a few generations after her, villages throughout the region had switched their way of life from one of hunting and gathering to one of herding goats, sheep and then cattle, and to growing domesticated crops. Selective breeding had transformed the plants and animals from their wild state to the service of humans within a remarkably short space of time. Sheep grew longer, woolly coats, which could be spun into garments. Goats provided a regular supply of milk. Cattle, domesticated from the ferocious wild aurochs, became docile suppliers of meat, milk and traction.

With food production and now the landscape increasingly under human control, the population increased relentlessly. This was partly due to a more consistent source of nutrition, but also because the new cereals, high in carbohydrates, removed the hormonal check on ovulation during lactation that had ensured a long gap between children. The increasing population was not all good news. It led to overcrowding and the arrival of epidemics of infectious disease which had never had a chance to take hold in the widely spaced bands of the hunter–gatherers. The close association of humans and animals after domestication enabled animal viruses, harmless to their hosts, to spread into the human population. Measles, tuberculosis and small-pox were caught from cattle, influenza and whooping cough from domesticated pigs and ducks. Judging from the signs of disease retained in their bones, the health of the early farmers suffered a sharp decline compared to their hunter–gatherer antecedents. Moreover, as people eventually abandoned hunting altogether and came to depend exclusively on a few crops and animals, they were vulnerable to famines when plants or animals failed due to drought or disease. But still the population grew. Nothing could stop the spread of farming. A thousand years after Jasmine, the unstoppable agricultural economy had crossed the Aegean from Anatolia and arrived in the plains of Thessaly in northern Greece. From the scarcity of hunter–gatherer archaeological sites of the same date in the region it looks as if this part of Europe was empty of humans at the time, until the farmers settled in. But elsewhere in Europe the hunter–gatherers were still doing well.

As the Great Ice Age ended, the southern edge of the tundra slowly receded. The rich game went with it, followed by the humans. The descendants of Ursula, Xenia, Helena, Velda, Tara and Katrine moved north to reclaim the great European plain. Behind them, the warmer climate encouraged the growth of trees and the landscape became one of thick deciduous forests with pines growing on the hills and mountains. Though not so productive as the tundra, these lands were still fully occupied by humans who looked increasingly to marine resources, fish and shellfish, to complement the reduced availability of game.

Old maps plot the spread of agriculture using large arrows curving across the surface of the globe with all the purpose of a well-planned military campaign. They show Europe embraced in a pincer movement from the bridgehead first established on the Greek mainland. On the southern flank, seaborne insurgents spread along the Adriatic and Mediterranean coast as far as Portugal. Meanwhile a massed attack on northern Europe was orchestrated from the Balkans as legions of farmers poured out of Hungary and occupied the continent from Belgium and France in the west to the Ukraine in the east. What hope did the locals have in face of this massive onslaught? But there was no such onslaught. Careful analysis of the archaeology of the early farming sites has certainly plotted the direction and timing of the spread of agriculture. These sites are easy enough to recognize, pottery and various agricultural implements, and the outline of huts in the ground, being among the obvious signs. But, as we saw with Jasmine's story, the whole essence of agriculture is that it can radiate quickly by word of mouth and by a few seeds and animals. It is an idea. It can spread. There is no need to insist that the spread of agriculture took the form of a large-scale invasion.

Recent archaeological investigations have shown that people took up farming at different rates in different places. The inhabitants of Denmark, for example, where the seafood harvest was rich enough to support a sedentary and prolific population, did not adopt agriculture on a large scale for over a thousand years after their neighbours only a hundred miles to the south. In other places, like Portugal, farming sites appeared not far from contemporary hunter–gatherer sites happily subsisting on the rich marine resources of the Tagus estuary. This does look like a new injection of people, probably only small in number, that brought the knowledge of farming by sea to new lands.

The new evidence from Europe which this book presents argues strongly in favour of our genetic roots being embedded firmly in the Upper Palaeolithic. Six of the seven women who are our ancestral mothers and whose imagined lives we have glimpsed were part of that resident population. They knew every inch of their landscape. They had good contacts with each other. They traded raw materials and finished goods. They were opportunists. If it suited them to farm, then they would take it up. It only needed someone to teach them; and among their tutors were the descendants of Jasmine. The mere fact that her descendants are alive and well and living in Europe is proof of the substantial genetic input from the Near East – substantial, but not overwhelming. Less than one-fifth of modern Europeans are in Jasmine's clan. The rest of us, with only a few exceptions, have deeper roots in Europe. At some time in the past our ancestors switched from hunting and foraging to embrace the farming economy. In more recent times some of the descendants of these ancestors abandoned the land for an urban existence sustained by the machine age. That is just another of the transformations that take place as people make individual decisions to take them to a better life.

Today, just under 17 per cent of native Europeans that we have sampled are in the clan of Jasmine. Unlike the other six clans, the descendants of Jasmine are not found evenly distributed throughout Europe. One distinctive branch follows the Mediterranean coast to Spain and Portugal, whence it has found its way to the west of Britain where it is particularly common in Cornwall, Wales and the west of Scotland. The other branch shadows the route through central Europe taken by the farmers who first cultivated the fertile river valleys and then the plains of northern Europe. Both branches live, even now, close to the routes mapped out by their farming ancestors as they made their way gradually into Europe from the Near East.

22

THE WORLD

The imagined lives of these seven women raise many questions. Were they the only women around at the time? We have seen very clearly that they were not. They lived and died among many other women. Ursula, for example, the oldest of our ancestral mothers, had many contemporaries. But she is the only one of them to whom a substantial proportion, about 11 per cent of modern Europeans, are connected by a direct maternal link. The maternal lines of her contemporaries did not make it through to the present day. At some point between then and now they petered out as women either had no children or had only sons. It is very likely that some of their genes which reside in the cell nucleus and which can swap between the sexes at every generation have made it through to today. But they will have arrived by a tortuous route which is impossible to trace. Many of Xenia's contemporaries, though not Xenia herself, would have been maternal descendants of the earlier Ursula. Likewise Helena, Velda, Tara and Katrine will have mixed with members of older clans. And when the descendants of

Jasmine arrived from the Near East with other agricultural pioneers, they would have passed on their knowledge to the descendants of the other six women.

Another frequently asked and reasonable question is whether there was anything special about these women, anything that would make them stand out from the others around them. Sadly, the answer is no – other than the necessary condition that each had to have two surviving daughters, there was probably nothing remarkable about them. They were not queens or empresses – such titles did not exist. They may or may not have been especially beautiful or heroic. They were essentially ordinary. Their lives were very different from ours today, but within their own time and people they would not have been exceptional. They had no idea they were to become clan mothers and feature in this book, just as any woman alive today with two daughters has the potential to be the founder of a clan which, were this book to be rewritten in fifty thousand years' time, might feature prominently on the cover. By then one or other of the seven clans may have drifted into extinction, to be replaced by others the founders of which are living somewhere today.

But perhaps the most intriguing enquiry is about the ancestors of the seven women themselves. Amazingly, we have also been able to discover the genealogy of these seven women. We can track back from the present day to reconstruct the mitochondrial DNA sequences of the seven clan mothers, then work out the ancestral relationship between them. I have retraced these connections in Figure 6. Each of the circles represents a particular

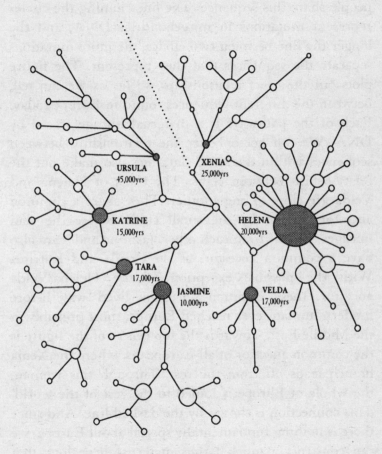

URSULA
45,000yrs

XENIA
25,000yrs

KATRINE
15,000yrs

HELENA
20,000yrs

TARA
17,000yrs

JASMINE
10,000yrs

VELDA
17,000yrs

Figure 6

mitochondrial DNA sequence, and the area of each circle is proportional to the number of people who share this sequence. The larger the circle, the more people share this sequence. The lines joining the circles represent mutations in mitochondrial DNA, and the longer the line between two circles, the more mutations separate the sequences that they represent. The figure plots out the exact relationships, so far as we can tell, between the different sequences found in Europe today. Each of the pathways is a maternal lineage traced by DNA. We can not only see the relationships between sequences within the same clan, but also make out the relationships between clans. The clans of Helena and Velda are close to one another. They share a common ancestor, shown by the small circle where the clan lineages split off from each other. Jasmine and Tara also have a common ancestor, as do Ursula and Katrine. With the possible exception of the Helena/Velda ancestor, these common ancestors lived way before modern humans ever reached Europe, most probably in the Middle East. Towards the top centre of the figure is the common ancestor of all Europeans, where the Xenia branch leads off from the rest. Through this woman, the whole of Europe is joined to the rest of the world. This connection is shown by the dashed line. And since there is nothing fundamentally special about Europe, we can construct a much larger maternal genealogy that embraces the entire globe.

Although most of this book has been about Europe, what I have described here can be done anywhere in the world. Over the last ten years active research

programmes have analysed and published mitochondrial DNA sequences from several thousand
people from all corners of the globe. We have put all of
these through the same process that we used to discover
the Seven Daughters of Eve. The end result of this
analysis is that we have discovered twenty-six other
clans of equivalent status in the rest of the world. About
some of these we know a lot; about others, very little.
Even so, I have given them all names. The picture will
no doubt change in the years to come as people from
previously unsampled regions volunteer their DNA.
But we know enough already to have a good idea and
to make a start on interpreting their meaning.

Of the thirty-three clans we recognize worldwide,
thirteen are from Africa. Many people have left Africa
over the last thousand years, a lot of them forcibly taken
as slaves to the Americas or to Europe. But their recent
genetic roots are quite clearly in Africa. Although
Africa has only 13 per cent of the world's population, it
lays claim to 40 per cent of the maternal clans. The
reason for this is that *Homo sapiens* has been in Africa
for a lot longer than anywhere else. The archaeology
supports this statement, the study of human fossils
supports it and now the genetics supports it too. There
has been a very long time for mutations to accumulate
in Africa. This means there has been time for new
clans to form and become distinctive and recognizably
different from one another. Different clans are more
frequent than others in some parts of the continent, but
there is no specific association between genetic clans
and tribal structures. This is a reflection of the great

antiquity of the genetic roots, which predate the formation of tribal and other classifications by more than a hundred thousand years.

Incredibly, even though the African clans are easily the most ancient in the world, we are still able to reconstruct the genetic relationships among them. Thus we probe the ancestors of the ancestors. At last, my dream of building a complete maternal genealogy for the whole of humanity was coming true. One by one the clans converge until there is only one ancestor, the mother of all of Africa and of the rest of the world. Her existence was predicted in the original scientific paper on mitochondrial DNA and human evolution in 1987. Immediately she was dubbed 'Mitochondrial Eve' – hardly a convincing African name. She lies at the root of all the maternal ancestries of every one of the six billion people in the world. We are all her direct maternal descendants. But, just as Ursula and the others were not the only women alive at the time, nor was Eve. Estimates of the size of the human population one hundred and fifty thousand years ago are bound to be not much more than guesswork, but it may have been in the order of one or two thousand individuals. Of these, only Eve's maternal lineage survives unbroken right through to the present day. The others withered away. But they, like Eve, would also have had maternal ancestors; so there is another woman even further back who was the ancestral mother of Eve and her contemporaries. She in turn would not have been alone, and another ancestral mother becomes a logical inevitability. This line of thought goes on and

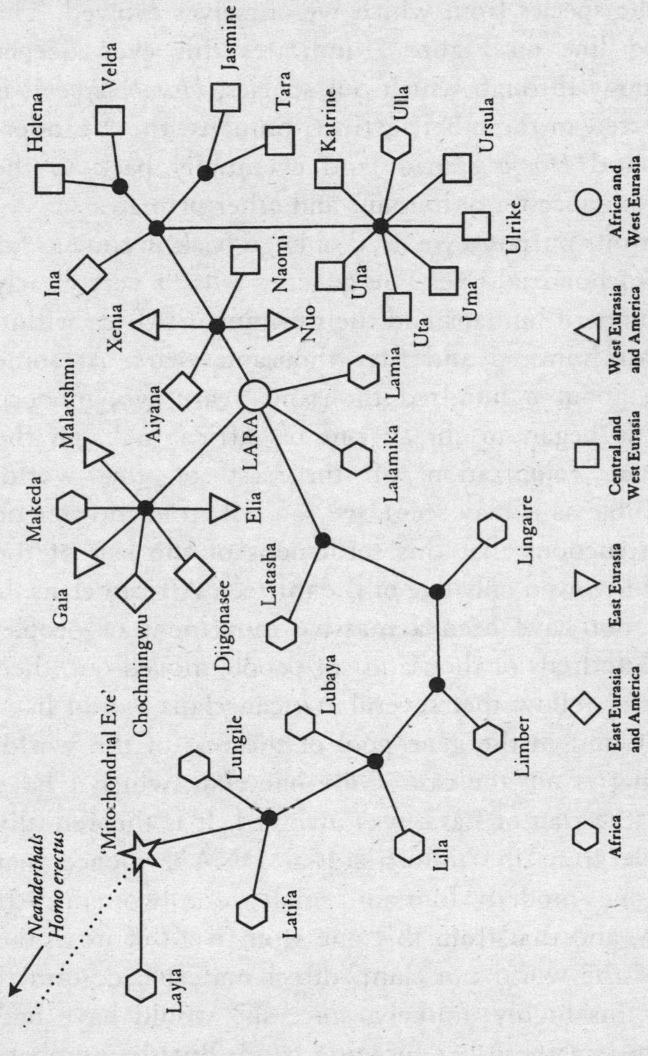

Figure 7

WORLD CLANS AND WHERE THEY ARE FOUND

Neanderthals
Homo erectus

'Mitochondrial Eve'

LARA

Layla

Latifa

Lila

Limber

Lubaya

Lungile

Chochmingwu

Djigonasee

Gaia

Makeda

Malaxshmi

Aiyana

Xenia

Ina

Helena

Velda

Jasmine

Tara

Katrine

Ulla

Ursula

Ulrike

Uma

Uta

Una

Naomi

Nuo

Elia

Latasha

Lalamika

Lamia

Lingaire

Africa

East Eurasia
and America

East Eurasia

West Eurasia
and America

Central and
West Eurasia

West Eurasia
and America

Africa and
West Eurasia

on, becoming increasingly pointless as we reach back millions of years to the very beginnings of our species and the species from which we ourselves evolved. The dashed line on Figure 7 indicates this even deeper genealogy through which our species, *Homo sapiens*, is connected to the other, extinct, humans, the Neanderthals and *Homo erectus*, and eventually back to the common ancestor of humans and other primates.

For our purposes we need only go back in time as far as Mitochondrial Eve. The genetics tells us very clearly that modern humans had their origins in Africa within the last hundred and fifty thousand years. At some point, about a hundred thousand years ago, modern humans began to spread out of Africa to begin the eventual colonization of the rest of the world. Incredible as it may seem, we can tell from the genetic reconstructions that this settlement of the rest of the world involved only one of the thirteen African clans. It could not have been a massive movement of people. Had hundreds or thousands of people moved out, then it would follow that several African clans would have been found in the gene pool of the rest of the world. But that is not the case. Only one clan, which I have called the clan of Lara, was involved. It is theoretically possible from the mitochondrial DNA evidence that only one modern human female, one woman, left Africa, and that from this one woman all of us in the rest of the world can claim direct maternal descent. I think this highly unlikely, since she would have had contemporaries in her foraging band. But the numbers must have been very small. This was no mass exodus.

Lara herself was not in the party. She probably lived in Kenya or Ethiopia; certainly in Africa. We know this because many Africans today are members of her clan. So she must have lived her life in Africa, unaware of her gift to the world, while it was her descendants that began to move out. Even so, it is a quite astonishing conclusion that the whole of the rest of the world can trace their maternal ancestry directly back to Lara. She is truly the mitochondrial Eve of the rest of the world.

All the evidence points to the Near East as the jumping-off point for the colonization of the rest of the world by modern humans. It was the only land route out of Africa, across Sinai. The only other possibility was to cross the Straits of Gibraltar at the entrance to the Mediterranean between north Africa and Spain. This is a deep channel which was never a land bridge, even when sea levels were at their lowest. Even so, the Straits of Gibraltar are only 15 kilometres across at their narrowest point, and the high Rock of Gibraltar is easily visible from the African side. But neither the archaeology nor the genetics suggest this route was taken.

There is good fossil evidence in Israel that *Homo sapiens* had reached the Near East at least one hundred thousand years ago. In this book we have traced the faltering spread of our species to the north and west into Europe, which finally succeeded only fifty thousand years ago. What held them up in the Near East for at least fifty thousand years before that? Europe was already inhabited by Neanderthals, physically adapted to the cold and experienced in the mechanics of making a living by hunting the large animals of the tundra. *Homo sapiens* in the Near East would have needed some advantage, however slight, over the Neanderthal occupants to make any headway. The long period spent in the Near East would have seen the slow developments in technology and, more important, in social interactions, that equipped them eventually to establish a permanent foothold in Europe.

The colonization of northern Asia was probably delayed for the same reasons. It too was a land dominated by steppe and tundra, running in a wide and uninterrupted ribbon from the Ukraine in the west to the high plateaux of Mongolia in the east. Archaeological sites in Mongolia dated to thirty-five thousand years ago witness the arrival of hunting bands with sophisticated flint arrow points in this bleak terrain at about the same time that modern humans were beginning to dominate the plains of western Europe. Their lives would have run along similar lines to the early Europeans we have already encountered, dominated by the seasonal migrations of the tundra animals and the fight to survive the unforgiving winters. We understand very little about the mitochondrial genetics of this vast region because it has not been widely sampled, but we do know enough to be able to be absolutely sure that it was from here that the colonization of the Americas was launched.

Four mitochondrial clans dominate the genetics of native Americans. All four have easily reconstructed and obvious genetic links with people living in Siberia or north–central Asia today. If they went by land, then their route into the Americas can only have been via Alaska. We have enough information about the sea-level changes over the past hundred thousand years to know that there were two periods when there was a continuous land bridge between Siberia and Alaska. The first bridge was formed fifty thousand years ago and lasted for about twelve thousand years. The second coincided with the last Great Ice Age, when the land was above sea level between twenty-five and thirteen thousand years ago.

There is a fierce controversy about when America was first colonized. Did the first people arrive across the earlier land bridge or the later one? There are two early archaeological sites in South America which have been used in the past to support the earlier date. One is at an open shelter at Pedro Furada in Brazil known for its rock painting. Flakes of paint have been found in the earth below the rocks at levels which have been dated to seventeen thousand years ago. But there is controversy about whether the flakes fell off the wall at that time or much more recently, working their way down into the lower levels through the action of worms or other creatures that disturbed the soil. The second site is at Monteverde in northern Chile, where fragments of wood, possibly part of a shelter, have been found at levels originally dated at thirty thousand years ago, although this has now been revised to a later date by the archaeologist who excavated the site. No human remains have been found at either Pedro Furada or Monteverde, and a big question hangs over the authenticity of both sites.

Perhaps the greatest evidence against the earlier date for the colonization of the Americas is that one would expect the population, in a land full of game and without prior human occupation, to explode, leaving abundant evidence all over the place. It is not as if nobody has looked. American archaeologists have worked hard to find it; but without success. However, there is plenty of evidence of a continuous settlement after twelve thousand years ago, with hundreds of sites scattered all across both North and South America.

The genetic evidence from modern native Americans also favours the later crossing. The accumulation of mutations in native Americans within each of the four clans has given all of them ages that fall well within the last thirteen thousand years. Reconstructions from modern Siberian and Mongolian patterns show very clearly that the clans were already established and separate from each other well before they reached America. The same applies to the rare fifth clan, that of Xenia, to which about 1 per cent of native Americans belong. As we have already seen, that clan had its origins on the borders of Europe and Asia.

The genetics fits well with the later land crossing from Siberia into western Alaska, just as the Ice Age was waning and the sea levels had begun to rise once again. But getting into Alaska was not the end of the story. Northern America was covered by two huge ice sheets. One enveloped the Rockies and the high mountains of southern Alaska; the other covered the whole of Canada. At the height of the last Ice Age, when sea levels were low enough to expose the land bridge from Siberia, these two great ice sheets fused to seal off access to the interior. The first Americans were faced with a dilemma. If it was cold enough to cross into Alaska by land, it was also too cold to get past the ice sheets on the other side. Alternatively, if it was warm enough to get through the ice sheets, by then the land bridge would be flooded. There had to be a period when the first Americans were stranded in western Alaska. Eventually the two ice sheets withdrew sufficiently to create a narrow corridor between them. This was no verdant valley, but a harsh passage though which the pioneers advanced little by little. At last the corridor opened out into the rich expanses of the Great Plains which were teeming with game. It must have been a wonderful and welcome sight to those first pioneers who had struggled through the ice corridor. From there on, the way was open for the rapid colonization of the whole of North and South America and, judging by the dates of the abundant archaeological sites, this was accomplished at record speed within only a thousand years.

The genetics agrees well with this scenario – except in one detail, namely that one of the four clans, the clan of Ina, is virtually absent from the modern inhabitants of Siberia and Alaska. It is found in South and Central America, and is still abundant in native Americans as far north as Vancouver Island on the north-west Pacific coast, but no further. Intriguingly, this same clan is also the one that is closely associated with the colonization of the Polynesian islands from south-east Asia. As we saw in an earlier chapter, the detailed sequences of

Polynesian and native American members of this widespread clan are sufficiently different to rule out a maritime colonization of the Americas from Asia directly across the Pacific via Polynesia. However, the curious absence of this clan from the present-day inhabitants of Siberia and Alaska suggests to me that we may be seeing the genetic echo of a second seaborne colonization that took the coastal route north up the coast of Asia and down the Pacific coast of North America. The rapid sea-level rises which flooded so much of south-east Asia would have given a great incentive to find new land. Could it be that the same maritime migration that ultimately led to the colonization of the remote Pacific islands also led a different branch of this remarkable clan to seek new land to the north – a journey which led them through polar waters and eventually to the temperate lands of Central America? What a voyage that would have been.

People from the Asian mainland also crossed to Japan at about the same time that they first reached America. One of the major questions in Japanese pre-history is the degree to which the modern population can trace its genetic roots to these earlier Jomon settlers, who are believed to have reached Japan about twelve thousand years ago, or to the much later Yayoi and subsequent migrations from Korea in the last two and a half thousand years. This issue has familiar parallels to the question of the composition of the gene pool of modern Europe and whether most Europeans trace their ancestry to the original hunter–gatherers or to the more recently arrived farmers from the Near East. We were able to settle this dispute using mitochondrial DNA. Could the same be done in Japan?

Comparatively little work has been done in Japan but there are hopeful signs that genetics will be able to decide this question. In addition to the Japanese on the central islands of Honshu, Shikoku and Kyushu, anthropologists recognize two other contemporary ethnic groups: the Ainu of Hokkaido in the north, and the Ryukyuans who live mainly on the southernmost island of Okinawa. One theory is that the Ainu and Ryukyuans are the descendants of the original Jomon settlers who occupied the whole of Japan and were then displaced from the central islands to Hokkaido in the north and Okinawa in the south by the arrival of the Yayoi from Korea. What little work has been carried out in Japan agrees in part with this idea, to the extent that it shows modern Japanese from the central islands sharing many more mitochondrial types with modern Koreans than do the Ainu and Ryukyuans. However, it also shows that the Ainu and Ryukyuans do not have very many shared types in common either. Age estimates, similar to the ones we did for the main clusters in Europe, show that both the Ainu and Ryukyuans have accumulated distinct mutations over the past twelve thousand years – which does suggest they are both the descendants of the original Jomon, but also that they have not been in close contact with each other since that time.

Although the majority of modern Japanese now live in Honshu, Shikoku and Kyushu, they do share many mitochondrial DNA sequences with modern-day

Koreans and so trace their maternal ancestry back to the Yayoi and subsequent migrations. Many other Japanese are also the maternal descendants of the Jomon and have their closest maternal relatives among the Ainu and the Ryukyuans. Whereas there is no doubt that the genetics confirms that the impact of the Yayoi settlers from mainland Asia was very substantial, far more so than that of the Near Eastern farmers in Europe, still it was not completely overwhelming. Much more needs to be done in Japan; but there is no doubt that mitochondrial DNA shows that modern Japanese are a mixture of Jomon and Yayoi and once again confirms that there is no such thing as a genetically pure classification into different races.

Both America and Japan were first reached by descendants of the hunting bands that had adapted to survive in the harsh conditions of the Asian tundra. This was a very different world from the one their ancestors knew in the Near East. It seemed to take about fifty thousand years spent in the Near East for *Homo sapiens* to acclimatize, both physically and organizationally, to these extreme conditions. But there was another way out of the Near East that did not involve adaptation to life on the tundra and an unrelenting diet of bison and reindeer. That exit was along the coasts of Arabia, the Persian Gulf and Pakistan, south of the great mountain ranges of central Asia, into India and south-east Asia. This route was much warmer, and much more like conditions in Africa, than the freezing northern route. It could have been used straightaway, without the long interlude adapting to the cold of the higher latitudes. Did people travel this southern route by sea thousands of years before their distant relatives eventually moved into Europe and northern Asia? Unfortunately there is no inland archaeology to support the idea of an ancient movement of people along this southern route, and thanks to sea-level rises coastal sites are now under water. But recently hand axes and flakes of the volcanic glass obsidian were found on a raised fossil beach at the edge of the Red Sea. Although no human skeletons were recovered from the site, which means we cannot be sure that the occupants were anatomically related to *Homo sapiens*, this is direct proof of human occupation of coastal sites at a very early date.

Whoever first discovered Australia certainly knew how to travel by boat. Even when sea levels were at their lowest, it was still necessary to make a journey of at least fifty kilometres over open sea to reach Australia. But how long ago did they arrive? Like the early American sites, the dating of very early archaeological finds in Australia has been controversial. However, judging by a recently dated burial site in south-east Australia, *Homo sapiens* was already there at least sixty thousand years ago. Even if these dates are only approximately accurate, they mean that modern humans reached Australia thousands of years before the colonization of Europe and northern Asia had even started.

If the archaeology is inconclusive, what can the genetics tell us? For understandable reasons, native Australians are very wary about participating in genetic tests, particularly those orchestrated by their former oppressors. The outcome is that only very few mito-chondrial sequences are known from native Australians. Those that have been published show only the remotest connection with the four clans from northern Asia that settled America. This rules out the possibility that the same hunters that crossed Asia north of the Himalayas and went on to colonize America also turned south and were the first to reach Australia. That much we can be sure of, and it does suggest that there might have been an earlier movement of people from the Near East across southern Asia. Sadly, we currently know so little about the mitochondrial genetics of native Australians that we are not in a position to be more specific about their genetic connections to people from other parts of southern Asia. From the few sequences that have been made public we can see that Australia probably holds several as yet unidentified clans. These are the signs of a very early arrival, leaving plenty of time for mutations to accumulate. They are also the signs of a relatively small population held constant over thousands of years. This fits well with what we know about the arid and hostile conditions that have persisted over this vast continent, which would have kept population growth to a minimum.

I am quite sure that genetics will be able to tell us a lot about how and when the first Australians arrived. I am equally sure that this history belongs to native Australians and not native Europeans like myself. It is their history, not mine. I, for one, would love them to share it with us.

23

A SENSE OF SELF

In the last chapter I could see myself slipping into the kind of language about human prehistory that I constantly try to avoid. It is the language of generalization, vitiated by the intentionality implicit in even such innocent-sounding phrases as 'the first Americans' or 'the first Australians'. There is an underlying suggestion that these were some sort of cohesive unit with an agreed policy – almost as if they had read the textbooks: 'OK, chaps, it's fifteen thousand years ago. Time to cross the Bering Land Bridge. And hurry up, it won't last for ever.' Even the Neanderthals: 'Sorry, lads. Time for us to go extinct and let the Cro-Magnons take over.' This is all complete and utter nonsense. There were no plans. How could there be? No-one can know what lies beyond the horizon. The whole of early human prehistory is based on the decisions of individuals or, at the very most, small bands of not more than a few dozen people.

There is solidity behind the statement: 'The Romans invaded Britain in AD 43.' That means something. A well-organized military empire can make decisions and

349

put in place large-scale actions to implement them. But this requires a far greater degree of organization and purpose than can ever have prevailed in our remote past. It is as if our present world of governments, corporations and committees has blinded us to the possibilities and importance of individual small-scale actions. I have tried to emphasize this in the imagined lives of the seven daughters. Though their whole existence depended entirely on uncontrollable elements of their environment – the movement of the herds, the advance and retreat of the ice caps – their day-to-day responses were a matter of individual choice within those constraints. In this view of human evolution, chance events and contingency are the variables. A boat sinks. A Polynesian island is not discovered for another hundred years.

I like this kind of genetics because it puts the emphasis back where it belongs: on individuals and their actions. This is much more appealing than the old-fashioned type of genetics, which was constrained by its methodology to force people into increasingly meaningless and misleading categories. Until I started this work I always thought of my ancestors, if I thought of them at all, as some sort of vague and amorphous collection of dead people with no solid connection to me or the modern world, and certainly no real relevance to either. It was interesting enough to read about what 'the Cro-Magnons' got up to all those years ago – but nothing much to do with me. But once I had realized, through the genetics, that one of my ancestors was actually there, taking part, it was no longer merely interesting – it is overwhelming. DNA is the messenger which illuminates that connection, handed down from generation to generation, carried, literally, in the bodies of my ancestors. Each message traces a journey through time and space, a journey made by the long lines that spring from the ancestral mothers. We will never know all the details of these journeys over thousands of years and thousands of miles, but we can at least imagine them.

I am on a stage. Before me, in the dim light, all the people who have ever lived are lined up, rank upon rank, stretching far into the distance. They make no sound that I can hear, but they are talking to each other. I have in my hand the end of the thread which connects me to my ancestral mother way at the back. I pull on the thread and one woman's face in every generation, feeling the tug, looks up at me. Their faces stand out from the crowd and they are illuminated by a strange light. These are my ancestors. I recognize my grandmother in the front row, but in the generations behind her the faces are unfamiliar to me. I look down the line. The women do not all look the same. Some are tall, some are short, some are beautiful, some are plain, some look wealthy, others poor. I want to ask them each in turn about their lives, their hopes and their disappointments, their joys and their sacrifices. I speak, but they cannot hear. Yet I feel a strong connection. These are all my mothers who passed this precious messenger from one to another through a thousand births, a thousand screams, a thousand embraces of a thousand new-born babies. The thread becomes an umbilical cord.

A thousand rows back stands Tara herself, the ancestral mother of my clan. She pulls on the cord. In the great throng a million ancestors sense the tug in lines that radiate out from her source. I feel the pull in my own stomach. On the bright stage of the living, I look to right and left and sense that others feel it too. These are the other people in the clan of Tara. We look at each other and sense our deep umbilical connection. I am looking at my brothers and sisters. Now I am aware who they are, I feel we have something very deep in common. I feel closer to these people than to the others. Like my ancestors, they are all very different to look at; but, unlike my ancestors, I can talk to them about it.

When two people find out that they are in the same clan they often experience this feeling of connection. Very few can put it into words, but it is most definitely there. Though DNA is the instrument which traces the links, I do not believe it has anything directly to do with the sensation. It seems inconceivable that the few genes which are embedded in the mitochondrial genome can directly influence feelings of this kind. They are certainly important genes and, as we saw in an earlier chapter, they allow cells to use oxygen. Without any evidence it would be hard to make a case that it was purely the similarities of cellular metabolism that caused this emotional feeling of shared experience. The DNA is certainly a physical object which has literally been passed from generation to generation, but its power is as a token or a symbol of the shared ancestry it reveals rather than the body chemistry it directly controls.

Many people experience a feeling of closeness and intimacy with others in the same clan. But would they feel this if the DNA tests had not revealed the connection? Two strangers enter a crowded room. Their eyes meet and they feel instinctively drawn to each other but don't know why. Are they acting under the influence of the subconscious awareness of an ancient connection? No research has yet explored this intriguing possibility, but as more and more people find out to which clan they belong, their reactions to their own ancestors and to each other will emerge.

What is it that we share with other members of our clan? We share the very same piece of DNA that has come down from our ancient maternal ancestors. We use it constantly. Cells in every tissue are reading the message it carries and carrying out its instructions millions of times a second. Every atom of oxygen we take into our bodies when we breathe has to be processed according to the formula that has been handed to us by our ancestors. That is a very fundamental connection in itself. But the route by which this gene reached us from those ancestors has its own special importance, for it follows the same path as the bond between a mother and her child. It is a living witness to the cycle of pain, nurture and enduring love which begins again every time a new child is born. It silently follows the mysterious essence of the feminine through a thousand generations. This is the deep magic which connects everyone in the same clan.

It is not a connection which is at all obvious in a world where family history and genealogy are dominated by inheritance through the male line. We are all familiar with the illuminated scrolls which celebrate the pedigrees of the rich and powerful. Without exception these trace the flow of titles, lands and wealth from father to son through the generations. Even the family trees of more modest households are built up around a scaffold of paternal inheritance. The immediate cause of this male monopoly on the past is simply that the written records on which all genealogy depends rely heavily on the use of surnames. With a surname as the only way into the records, it is no surprise that what comes out the other end is a family tree drawn around men. But the ultimate cause is the same patriarchal attitude of Western civilization that we encountered in the first theories of inheritance. Wealth and status were the only things considered to be worth inheriting, and they passed down the male line.

The common practice of women adopting the husband's surname on marriage rather than retaining their maiden names makes it very difficult to trace a maternal lineage, for women's names change at every generation. But neither would retention of the maiden name resolve the problem, because a maiden name is, after all, only another surname – a father's name rather than a husband's. Against this background it is no surprise that it comes as a revelation to many people that there actually is such a thing as a maternal family tree, a mirror image of the traditional paternal version. I have certainly never seen one drawn out.

Genetics does help to reconstruct detailed maternal trees even within the existing records, but the best solution for future generations of genealogists would be to create a new class of name altogether. Everyone would get this name from his or her mother. Women would pass it on to their children. It would be, in effect, an exact mirror image of the present system with its surnames which people get from their fathers and, if they are men, pass on to their children. We would then all have three names: a first name, a surname and a new one, a *matriname* perhaps. A man passes on his surname to his children; a woman gives her matriname to hers. Since they follow a maternal line of inheritance, these names will closely correspond with mitochondrial DNA. They will also reflect biological relationships more accurately than surnames, because there is only very rarely any doubt about the identity of a child's mother. In time people would be able to recognize their maternal relatives with the same matriname in just the same way as they can now link up to their paternal family through a shared surname. But until that time comes, if it ever does, reconstructing maternal family trees through written records alone will be much harder than drawing the male equivalent.

In the short time during which I have been able to help people reconnect to their ancestors or their relatives using DNA, I have received many requests from individuals who have tried to establish a link through the records but for one reason or another have not been able to do so. Paper records can be destroyed by fire, eaten by termites, erased by mould or simply just lost. DNA is able to fill in the gaps created by missing records. This helps to compensate for the inherent frailty of pen and paper; but there are many people for whom the lack of any written records about their ancestors is not an accident but is deliberate obliteration. In these cases, DNA is not just a useful supplement to the traditional techniques of the genealogist. It becomes their only physical link to the past.

For Jendayi Serwah, establishing a link to her past was a mission of great personal significance. She is a lady from Bristol whose parents had each arrived in Britain from Jamaica as teenagers. Their ancestors had been taken from Africa as slaves to work the plantations. But there were no records of this. The only details the slave ships kept were the most basic description of their human cargo: how many men and how many women were loaded on board, and how many survived the long sea voyage, was all that was written down. After they were landed and sold on to the plantation owners, their individuality was deliberately erased. They were given European names. No records were kept of births or marriages or deaths. Their pasts as individuals were intentionally obliterated. It was not that it would have been difficult for Jendayi to trace her ancestors in Jamaica back more than a few generations; it would have been completely impossible. Of course, she guessed that her deep ancestry lay in Africa; but there was no real proof of it, other than the general historical presumption that many captives from west Africa were sold to plantation owners in the Caribbean. So it was not surprising that, when we tested her DNA, Jendayi had a mitochondrial signature that was clearly African. However, when I told her of this result and also that we had found a very close DNA match with a Kenyan Kikuyu, the effect on her was overwhelming. She was literally lost for words. Here at last was the *individual* proof she had wanted for so long. It was as if the DNA was itself a written document from her ancestors, which in a sense it was; a document that had been handed down, one generation at a time, from the woman who had endured and survived that terrible voyage from Africa. A document that could not be obliterated by the plantation owners as it passed unseen and unread through the generations. And now in Jendayi here it was, a perfect copy of the African original preserved within her own body.

I have seen many other astonishing journeys witnessed by this remarkable piece of DNA. In western Europe more than 95 per cent of native Europeans fit easily within one or other of the seven clans. That still leaves a large number of people whose deep maternal lineages tell of a different history. Unlike Jendayi, they are usually completely unaware of the exotic journeys recorded in their DNA. For instance, a primary school teacher from Edinburgh carries the unmistakable signature of the Polynesian mitochondrial DNA which I can recognize from a mile off. She knows her own family history well for the past two hundred years, and there is nothing that gives any clue as to how this exotic piece of DNA came to her from the other side of the world. But there is no doubt that it did. What tales it could tell of the South Seas! Is she perhaps the descendant of a Tahitian princess who fell in love with a handsome ship's captain, or of a slave captured by the

Arabs on the coast of Madagascar? There are many other equally mysterious journeys recorded in our DNA: the Korean sequence that turns up regularly in fishermen from Norway and northern Scotland; the unmistakably African DNA in a dairy farmer from Somerset, a legacy perhaps of Roman slaves from nearby Bath; the sequence of a book salesman from Manchester that is so unusual that his closest match is found among the native Australians of Queensland.

One outstanding genetic journey involves a complete circumnavigation of the world. Two fishermen on a small island off the west coast of Scotland have unusual mitochondrial sequences, and I thought at first they might be closely related to one another, although they had no knowledge of it. As we discovered more sequences from different parts of Europe and the rest of the world, we began to find much closer matches to the two men – one in Portugal and one in Finland. These were still unusual sequences to find in Europe, not part of the seven original clans. The Portuguese sequence matched several from South America, and the Finnish DNA was close to sequences found in Siberia, where we also found the ancestral sequence of the South Americans. So the two fishermen were indeed related – but only through a common ancestor from Siberia. One line of maternal ancestors had travelled from Siberia along the coast of the Arctic Ocean to Scandinavia, then on to the west of Scotland, perhaps aboard a Viking ship. The other had crossed into America over the Bering Straits, then down to Brazil. At some time, presumably after Brazil became a Portuguese colony, a woman carrying this piece of DNA crossed the Atlantic to Portugal, from where, somehow, it had found its way up the Atlantic coast to the west coast of Scotland. The two journeys had ended on the same small island after travelling in opposite directions from the other side of the world.

These stories and others like them make nonsense of any biological basis for racial classifications. What I have related here is only the tip of the iceberg, the clear message from the gene that is the easiest to read. The tens of thousands of other genes in the cell nucleus would echo the same message. We are all a complete mixture; yet at the same time, we are all related. Each gene can trace its own journey to a different common ancestor. This is a quite extraordinary legacy that we have all inherited from the people who lived before us. Our genes did not just appear when we were born. They have been carried to us by millions of individual lives over thousands of generations.

At a recent conference I sat aghast in the audience as patent lawyers and biotechnologists debated the pros and cons of patenting genes. The arguments were legalistic in the extreme. DNA, to the lawyers, was just a chemical. Since it could be artificially synthesized, they argued, why should it not be patented like any other chemical? At one point an enthusiastic manager from a large pharmaceutical company stood up to address the audience. He was summarizing the current position, and illustrated his point with a pie-chart showing the division of ownership of the human genome, the sum total of all human genes, among major corporations. The pie was sliced up and the portions assigned. The financial arguments were impeccable. You could not expect major investment by pharmaceutical companies into genetics unless these investments could be protected by patents. Patents are being filed every day claiming ownership and a commercial monopoly on our genes. As I sat there, I had the overwhelming and very disturbing sensation that parts of me and my past were being bought and sold.

As the presentation continued I reflected on the fact that I was sitting here, in a conference room, at one of the most advanced DNA facilities in the world, while in vast halls on either side, rank upon rank of robotic machines were silently reading the secrets of the genome. An electronic board in the lobby continuously flashed up the DNA sequences as they came off the machines. Before my very eyes the details of the genome that had been hidden for the whole of evolution were marching across the screen. Was this, the reduction of the human condition to a string of chemical letters, the ultimate expression of the Age of Reason that first began to separate our minds from our intuition and to distance us from nature and our ancestors? How ironic that DNA should also be the very instrument that reconnects us to the mysteries of our deep past and enhances rather than diminishes our sense of self.

Not 'just a chemical' after all, but the most precious of gifts.

INDEX

Abingdon cemetery, 29-30, 32, 33, 214

Africa, human evolution:
debate, 71-2
fossil record, 142-3, 335
genetic evidence, 335-9
Homo sapiens, 165
movement out of, 146
technology, 147
slaves from, 335, 356

agriculture:
beginnings, 168-70, 213-14, 323-6
development, 185-7, 327-9
health effects, 171, 327

Ainu, 345

Alaska, land bridge, 341-4

Alexandra, Tsarina, 86, 88, 89

Alexei, Tsarevich, 89, 100-101

America, colonization of, 341-4, 346

American Journal of Human Genetics, 192, 193

Americans, native, 340-1

amino-acids, 27, 47, 48

Ammerman, Albert, 186-7

Anderson, Anna, 97-100

animals, domestication of, 172, 325-6

Aristotle, 39, 40

art:
carvings, 293-4, 297, 299
cave paintings, 147, 281-3, 293

Asia, 165, 340

Australia, 165, 169, 347-8

Avdonin, Aleksandr, 86

Bandelt, Hans-Jürgen, 177-8

Barcelona, Second Euroconference on Population History, 190-2